Sliding Window Algorithm for Mobile Communication Networks

Nuka Mallikharjuna Rao
Mannava Muniratnam Naidu

Sliding Window Algorithm for Mobile Communication Networks

 Springer

Nuka Mallikharjuna Rao
Department of Master of Computer
  Applications
Annamacharya Institute of Technology
  and Sciences (Autonomous)
Rajampet, Andhra Pradesh
India

Mannava Muniratnam Naidu
School of Computing
Vel Tech Rangarajan Dr. Sagunthala R&D
  Institute of Science and Technology
  (Deemed to be University Estd. u/s 3
  of UGC Act, 1956)
Chennai, Tamil Nadu
India

ISBN 978-981-10-8472-0   ISBN 978-981-10-8473-7   (eBook)
https://doi.org/10.1007/978-981-10-8473-7

Library of Congress Control Number: 2018933460

Printed on acid-free paper

This Springer imprint is published by Springer Nature
The registered company is Springer Nature Singapore Pte Ltd.
The registered company address is: 152 Beach Road, #21-01/04 Gateway East, Singapore 189721, Singapore

Preface

The primary objective of this book is to discuss how to improve the throughput of Mobile Switching Center (MSC) in Global System for Mobile Communications (GSM) network.

The book begins by building the core knowledge of Global System for Mobile Communications in Chapter 'Introduction.' It presents an overview of GSM network architecture and databases. Many approaches relevant to prior work are discussed.

Chapter 'Sliding Window Algorithm' discusses fixed block of seven days algorithm and develops an approach for maximizing GSM network throughput and minimizing the call setup time by using proposed sliding window of size seven days algorithm. The key advantage of this model is to reduce call setup time between the caller and the receiver in a network.

Chapter 'Performance Measurement of Sliding Window Algorithm' discusses a simulation model for evaluating the performance of fixed block of seven days and sliding window of size seven days algorithms. Chapter 'A Model for Determining Optimal Sliding Window Size' discusses a model to determine optimal window size in order to maximize the network throughput and minimize call setup time.

Chapter 'Integrating Sliding Window Algorithm with a Single Server Finite Queuing Model' presents integration of sliding window algorithm with a single server finite queuing model. Then, a simulation model is developed for evaluating the performance of sliding window of size seven days algorithm and integrated model (IM) at an MSC service area employing call setup time and throughput as performance criterion.

Chapter 'Integrating Sliding Window Algorithm with a Multiple Server Finite Queuing Model' presents integration of sliding window algorithm with a multiple finite queuing model. Later, a simulation model is developed for evaluating the performance of sliding window of size seven days algorithm and integrated model with a multiple channel (IMMC) at an MSC service area with regard to call setup time and throughput. Through simulation results, there is a significant increase in performance metrics of the proposed integrated model (IM) and IMMC for one

MSC service area. Obviously, it is recommended to consider adopting IM and IMMC for the entire GSM network for improving its throughput by 4.78%.

Chapter 'Method for Determining Optimal Number of Channels' discusses a decision model for determining the optimal number of channels. Average call setup waiting time in system and idleness percentage of channels are used as criteria of optimization. A simulation model is formulated by employing the aspiration decision model for profiling the behaviour of average call setup waiting time in system and idleness percentage of channel as a function of number of channels. It is employed to simulate assuming sliding window of size seven days. It is found that as the number of channels increases, the average call setup waiting time in system decreases and idleness percentage of channels increases. It facilitates the decision maker to choose the optimal number of channels for the chosen aspiration/service levels.

Rajampet, India   Nuka Mallikharjuna Rao
Chennai, India   Mannava Muniratnam Naidu

Acknowledgements

It is not surprising that this momentous time of my life would have been impossible without the support, enthusiasm, and encouragement of many incredibly precious people. Hence, I dedicate this preamble to them.

First and foremost, I would like to thank 'Dr. Mannava Muniratnam Naidu' for giving me the opportunity to work with him and under his supervision as co-author. I am very much grateful to him for his invaluable guidance and insightful comments on my book and the discussions which I had with him and also for his concern about many other things which are not related to the work. I am sure I would not have been able to finish this book without his help and remarkable ideas concerning the publications that I co-authored with him. For all this and more, I gratefully thank him.

I am grateful beyond expression to my dearest family. I feel that now, at the end of this work, is the relevant time to express my best thanks to them for their unconditional support, encouragement, and faith in me throughout my whole life, in particular during the last four months. I hope that I will be able to compensate them in the future. I dedicate this book to them, with love and gratitude.

Contents

Introduction 1
1 Introduction 1
   1.1 Types of Mobility 1
   1.2 Mobility Management 3
   1.3 Roaming 5
   1.4 GSM Network Architecture 6
   1.5 Models and Paradigms 11
2 Summary 13

Sliding Window Algorithm 15
1 Introduction 15
   1.1 The Model 15
2 Fixed Block of Seven Days (FBSD) Algorithm 17
   2.1 Method 17
   2.2 Illustrative Example 19
3 Sliding Window of Size Seven Days Algorithm 23
   3.1 Method 24
   3.2 Algorithm 24
   3.3 Algorithm Design Steps 29
   3.4 Illustrative Example 31
4 Summary 34

Performance Measurement of Sliding Window Algorithm 35
1 Introduction 35
2 Performance Metrics 35
   2.1 Hit Rate 35
   2.2 Throughput 36
3 Simulation Model 36
   3.1 Simulation Parameters 38
4 Experimentation 40

5   Simulation Output Analysis   40
6   Summary   52

A Model for Determining Optimal Sliding Window Size   55
1   Introduction   55
2   The Model   55
3   Simulation Process   57
4   Simulation Output Analysis   60
5   Summary   66

Integrating Sliding Window Algorithm with a Single Server
Finite Queuing Model   67
1   Introduction   67
2   A Single Server Finite Queuing Model   67
3   Integration of Sliding Window Algorithm with a Single
    Server Finite Queuing Model   68
4   Simulation   69
5   Summary   75

Integrating Sliding Window Algorithm with a Multiple Server
Finite Queuing Model   77
1   Introduction   77
2   Multiple Channel Finite Queuing Model   77
3   Integrating Sliding Window Algorithm with a Multiple
    Channel Finite Queuing Model   78
4   Simulation Output Analysis   79
5   Summary   85

Method for Determining Optimal Number of Channels   87
1   Introduction   87
2   The Model   88
3   Simulation Process   91
4   Summary   92

Glossary of Abbreviations   93

References   95

About the Authors

Nuka Mallikharjuna Rao received his B.Sc. in Computer Science from Andhra University, Visakhapatnam, Andhra Pradesh, India, in 2005 and MCA in Computer Applications and Ph.D. in Computer Science and Engineering from Acharya Nagarjuna University, Guntur, in 2008 and 2015, respectively. He is a Life Member of the Indian Society for Technical Education (ISTE) and a Member of IEEE, IACSIT, and CSTA. He is presently working as a Professor of Computer Applications and Director of the Internal Quality Assurance Cell (IQAC) at the Annamacharya Institute of Technology and Sciences, Rajampet. He has more than 18 years of teaching experience, and his current research interests include mobile computing, mobile networks, distributed networks, and queuing theory.

Mannava Muniratnam Naidu received his B.E. in Mechanical Engineering from Sri Venkateswara (SV) University, Tirupati, and master's degree in Engineering and Ph.D. from the Indian Institute of Technology Delhi (IIT Delhi), Delhi, India. He served as a convener and member of many committees on behalf of the All India Council for Technical Education (AICTE). He is a Life Member of ISTE, ORSI, ISME, CSI, IEEE, and ACM. He served as a Professor, Dean, and Principal at the SV University College of Engineering, Tirupati. He also worked as a Professor in the Department of Computer Science and Engineering, Vignan University, Guntur, Andhra Pradesh. Currently, he is working as the Dean of Computing at Vel Tech University, Avadi, Chennai, India. His research interests include data mining, computer networks, soft computing techniques, and performance evaluation for algorithms.

List of Figures

Introduction

Fig. 1   Location management operations......................... 3
Fig. 2   GSM network architecture............................... 6
Fig. 3   BSS system architecture 7
Fig. 4   Process flow between MSC, HLR, and VLR................. 9

Sliding Window Algorithm

Fig. 1   Activity diagram for naïve method 16
Fig. 2   Activity diagram for FBSD first-day call setup requests 20
Fig. 3   Activity diagram for FBSD algorithm 21
Fig. 4   **a** First set of seven days block with no records, **b** first set
of seven days block with subscriber records, **c** after first
intersection process, **d** next seven sets block with records....... 22
Fig. 5   Activity diagram for first 'n' days......................... 30
Fig. 6   Activity diagram for subsequent days 31
Fig. 7   **a** First window with seven days with day 1 records, **b** first
sliding window with seven days' records, **c** after first slides
of window, **d** after sliding window slide right by one day,
e window slide right by another day, **f** after sliding window
slide right by one day 32

Performance Measurement of Sliding Window Algorithm

Fig. 1   Simulation model.. 37
Fig. 2   Sample Poisson random variates 39
Fig. 3   Sample discrete random variates for a given day............. 39
Fig. 4   ISPF 6.0 editor 40
Fig. 5   IBM programming interface............................. 41
Fig. 6   JCL environment....................................... 41
Fig. 7   Input data view 42
Fig. 8   DB2 database environment IBM technologies............... 42

Fig. 9   System-generated output for FBSD algorithm   43
Fig. 10   System-generated output for SWSSD algorithm   43
Fig. 11   Hit rate versus blocks at an MSC   47
Fig. 12   Throughput versus blocks at an MSC   48

A Model for Determining Optimal Sliding Window Size

Fig. 1   Sliding window size versus AHT, AMT, and ACST   63
Fig. 2   Sliding window size versus average call setup time over
            a range of 7–15   64
Fig. 3   Sliding window size versus throughput over a range of 7–15   64
Fig. 4   Sliding window size versus average call setup time over
            a range of 7–49   65
Fig. 5   Sliding window size versus throughput over a range of 7–49   65

**Integrating Sliding Window Algorithm with a Single Server
Finite Queuing Model**

Fig. 1   Integrated model   69
Fig. 2   Waiting time in system over 1001 days.   73
Fig. 3   Waiting time in queue over 1001 days   73
Fig. 4   Throughput over 1001 days.   74
Fig. 5   Call setup time in system versus blocks at a MSC   75
Fig. 6   Throughput versus blocks at a MSC   75

**Integrating Sliding Window Algorithm with a Multiple Server
Finite Queuing Model**

Fig. 1   Integrated model with a multiple channel   78
Fig. 2   Average waiting time in system versus blocks at a MSC   84
Fig. 3   Throughput versus blocks at a MSC   84

Method for Determining Optimal Number of Channels

Fig. 1   Number of channels versus average waiting time in system   92

List of Tables

Performance Measurement of Sliding Window Algorithm

Table 1 Simulation input parameters 38
Table 2 Input and output of simulation for each block 44
Table 3 Computing of performance metrics 48
Table 4 Comparison of performance metrics 52

A Model for Determining Optimal Sliding Window Size

Table 1 Simulation parameter values 57
Table 2 Performance metrics versus sliding window size
         for simple indexed sequential file 60
Table 3 Performance metrics versus sliding window size
         for multi-level indexed file 62

**Integrating Sliding Window Algorithm with a Single Server
Finite Queuing Model**

Table 1 Simulation parameters 70
Table 2 Simulation results 71
Table 3 Aggregated simulation results for each block 74

**Integrating Sliding Window Algorithm with a Multiple Server
Finite Queuing Model**

Table 1 Input/output simulation parameters 80
Table 2 Simulation results 81
Table 3 Aggregated simulation results for each block 83
Table 4 Comparison of performance metrics 85

Method for Determining Optimal Number of Channels

Table 1 Simulation parameters 91
Table 2 Average waiting time with channel idleness 91

Introduction

1 Introduction

The word 'mobile' has completely changed the world of communications giving scope for origination of innovative applications that are limited to one's imagination. In present days, cellular communication has become the backbone of the industry and society. All the mobile communication technologies have improved the way of living in the society. The Global System for Mobile (GSM) communications is an extraordinary stage of successful development of modern information technology. Currently, more than 900 million users subscribed GSM networks, and this number can increase exponentially in future. Over 120 countries deployed GSM services to improve the utility of application pertaining to individual and corporate sectors. This is a vital technology that provides communication services for the rapid subscriber growth in mobility. The GSM comes across the challenges in providing services for subscribers with mobility. This book presents a new approach in mobile communications called sliding window algorithm for improving the throughput of Mobile Switching Center's (MSCs) in GSM network. The gap between naïve and current mobility techniques and the vision for future mobility for location identity indicate that much work remains to be done to make this vision a reality.

1.1 Types of Mobility

Mobility or mobility management is a functionality that facilitates mobile user operations in GSM networks. Mobility is used to trace the geographical area of user and user locations to provide mobile phone services for making calls and transferring data between users.

© Springer Nature Singapore Pte Ltd. 2017
N. Mallikharjuna Rao and M. Muniratnam Naidu, *Sliding Window Algorithm for Mobile Communication Networks*, https://doi.org/10.1007/978-981-10-8473-7_1

The main characteristic and purpose of mobility are to identify subscribers, wherever they are allowed to make calls and to deliver mobile services to them. In general, mobility is classified in two ways: *terminal mobility* and *personal mobility* [1].

The wired, wireless components and subscribers (humans) are the main components in a mobile framework. The system exists two other important parts, namely terminal mobility and personal mobility. These parts are eliminating spatial and temporal constraints from the call setup and data processing activities on wired and wireless devices.

GSM network is to provide terminal mobility, while the terminal is roaming to access the telecommunications service from different locations and capacity of the GSM network to monitor the terminal. The terminal mobility supports to connect any mobile user from anywhere by any mobile user.

The personal mobility supports to the mobile user that mobile user does not require to hold any equipment with user for communication and establishing communication with other mobile user.

A mobile user wishes to communicate with other user; a verification process is required in order to authenticate concerned user that is done through an identification scheme. Personal mobility, enabling the mobile user, can use services to connect to the network or other specific terminal.

Now days, personal mobility is creating many new challenges in telecommunication networks, since it is well known in the network where potential subscribers are. The realistic development is based on the measurement of location and possibly mobility models are time-dependent where subscribers are greatly facilitated with the design of cost-effective network which also meets the demands of the subscribers. Terminal and Personal mobility is independent from each other and can exist without user. The personal mobility is the ability of a user to access telecommunication services at any terminal on the basis of a personal identifier, and the capability of the network to provide those services in accord with the user's service profile where as the terminal mobility, it is necessary to associate service subscription with the terminal itself. Both the mobilities are support to voice and data communication. However, both types of mobility techniques are essential to visualize a complete database system of a mobile device.

To manage the records (profiles) of subscribers, public land mobile network (PLMN) has several databases. It has identified two units of databases for subscriber registration and current position as home location register and visitor location register. This organization is based on the number of subscribers, the capacity of processing and storage, and switches on the network structure. Mobile databases are used to store calling information made by subscribers across the network. It is also used for monitoring the subscriber information and for identifying the present location of mobile user.

1.2 Mobility Management

The important aspect and challenging problem in mobility management are providing seamless mobile accessing service. The essential technology used in it is to automatically support mobile terminals enjoying their hassle-free seamless roaming services exclusive of drops in communication. The key aspects to be considered in mobility management are location management and handoff management. This book is mainly focused on location management and its process as we shown in Fig. 1.

In wireless networks, mobile users are enjoying the services for making and receiving calls when they are in roaming which is referred as mobility management and it is fundamental technology that enables mobile users roaming across different locations.

Mobility management is the fundamental technology that enables users to roam with their mobile terminals to enjoy the services in progress through wireless networks. From the point of view, the functionality of mobility management enables communication networks to do perform the following:

- In order to deliver data packets, the system locates roaming devices frequently, i.e., static approach.
- Usually, mobile user roams randomly between home network to visited network, the location identities of roaming users are necessary for providing services to their call setup requests. Hence, the network has to maintain user movements and its new connections dynamically, i.e., dynamic apporach.

1.2.1 Location Management

In mobile communications [2], a mobile is a unit which moves across the network converge area freely at any time and any place. The movement of the mobile user is random, and hence, their geographical location is unpredictable which makes it

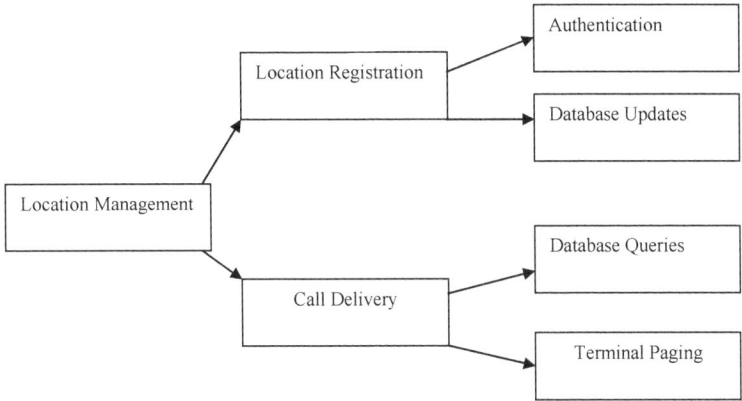

Fig. 1 Location management operations

mandate to locate the cell and record its location to home location register (HLR) and visitor location register (VLR) when a call has to be delivered to it.

The location management is kind of directory management in which current locations are maintained continuously. The objective of location management approach is to reduce the communication overhead owing to number of database updates (i.e., HLR interactions) [3].

The fundamental tasks of location management are:

1. Location lookup
2. Location update
3. Paging.

The location update operations are initiated by the mobile device, and the current location of the device is stored in home location register and visitor location register. The location lookup is a process to identify the present cell location of mobile user by fetching pertinent record from home location register and inform caller present location to called device about its present serving base station. The location update and location lookup operations are initiated by a Mobile Switching Center. As size of cell is low, the cost of record updating and paging automatically becomes high. Further, cost is increased whenever mobile user crosses high commuting zones frequently. The objective of initiating location areas and paging areas is to reduce the cost of network. A group of neighboring cells is formed as a group, it referred as location area (LA), and the paging areas are also formed same way.

In some occurrences, some of the remote cells are also be included in the locate areas. For as long as the mobile device moves intra-cells of a location area, location update operation is not essential because mobile user identity is available in its neighboring cells. It minimizes the cost of network because the location updates as well as paging operation are not performing every time.

There is another issue which is related to the distribution of HLR in order to shorten the access path, a similar kind of approach in data distribution problem in distributed database systems [5]. These prominent issues motivated many authors to present a number of innovative research articles for location management schemes.

The current location of mobile user is articulated in terms of the cells or the base station to which device is currently connected. The mobile device (called and calling mobile users) can moves randomly around or near or within their intra-base stations and thier identity verification is not essential. Usually, the location registration, lookup operations are invoked to identify the new location of mobile user when they are exit when they roams between inter-base stations.

The arrivals and departures of subscribers related to a MSC service area are random in nature. Whenever a subscriber arrives at a MSC service area, the VLR is searched for relevant record (i.e., subscriber profile which contains complete information about mobile subscriber). If the record is found, the call set-up is made. Otherwise, the subscriber's profile is fetched from the HLR to facilitate call setup and it is inserted into the VLR and it causes high network traffic overhead by which call setup is delayed. In another approach, is replicates HLRs data at each VLR, which helps in reducing network overhead and time for call setup. However, it increases the storage space requirements enormously and relevant record access time.

A Mobile Switching Center establishes call connection in response to the call setup request received from a mobile user through pertinent base transceiver station and base station controller. Whenever a Mobile Switching Center receives a call setup request from a mobile user in its location service area, it cannot establish connection unless its visitor location register contains mobile user data record. Mobile user roams among location service areas randomly. Generally, the data record of a mobile subscriber is deleted from visitor location register whenever he leaves the current location service area. It is fetched from home location register whenever the mobile user reenters the location service area, which would result in network latency leading to increase in call setup time. The network latency can be eliminated, provided the data record of mobile user is retained even though the mobile user leaves the location service area. However, it increases visitor location register storage space requirements and records access time. Then, the problem is to formulate a policy for determining the retention period of the data record of a mobile subscriber in visitor location register, in spite of its exit from the current location service area. Such a policy shall strike a balance between call setup time and visitor location register storage space requirements.

Newaz Shah proposed a method of fixed block of seven days (FBSD) for determining the retention period of the data record of a mobile subscriber in visitor location register for improving the throughput of the Mobile Switching Center. Nuka and Naidu proposed an algorithm, sliding window of size seven days (SWSSD). They proved through simulation study that the performance of SWSSD is better than that of FBSD based on average call setup time or equivalent throughput. Further, Nuka and Naidu proposed a model for determining the optimal sliding window size (OSWS) augmenting incremental visitor location register access time which is function of visitor location register size.

Nevertheless, they assumed that the waiting time of call setup requests in queue for want of a Mobile Switching Center time as insignificant. It is far from the real-life situation wherein the waiting time of call setup requests in queue is significant.

1.3 Roaming

Roaming is the facility of accessing seamless connectivity with the help of many service providers. Thus, when a mobile users moves from one service provider to any other service provider/network over GSM or any other network, the location of mobile unit must be informed by the new service provider to the old service provider. This facility is called roaming facility.

In location management, a service provider communicates with each other to complete registration process which is described earlier. The other notable aspect while subscriber is in roaming is billing as subscribers move among multiple network providers who offer services at different prices.

1.4 GSM Network Architecture

The Global System for Mobile (GSM) communications network facilitates the communication between geographically separated mobile subscribers using their mobile wireless mobile stations with subscriber identification modules (SIM) cards.

This allows mobile subscribers to roam all over the world. Its integration with the network of integrated data services (ISDN) is simple. It offers a service quality ensuring high security. It handles high volume of calls offering more channels with improving limited bandwidth spectrum efficiency. The GSM network architecture that is presented in [6, 7] is elaborated and shown in Fig. 2.

The international mobile equipment identity (IEMI) is identification number of a mobile station, which provides mobility and is referred to as terminal mobility.

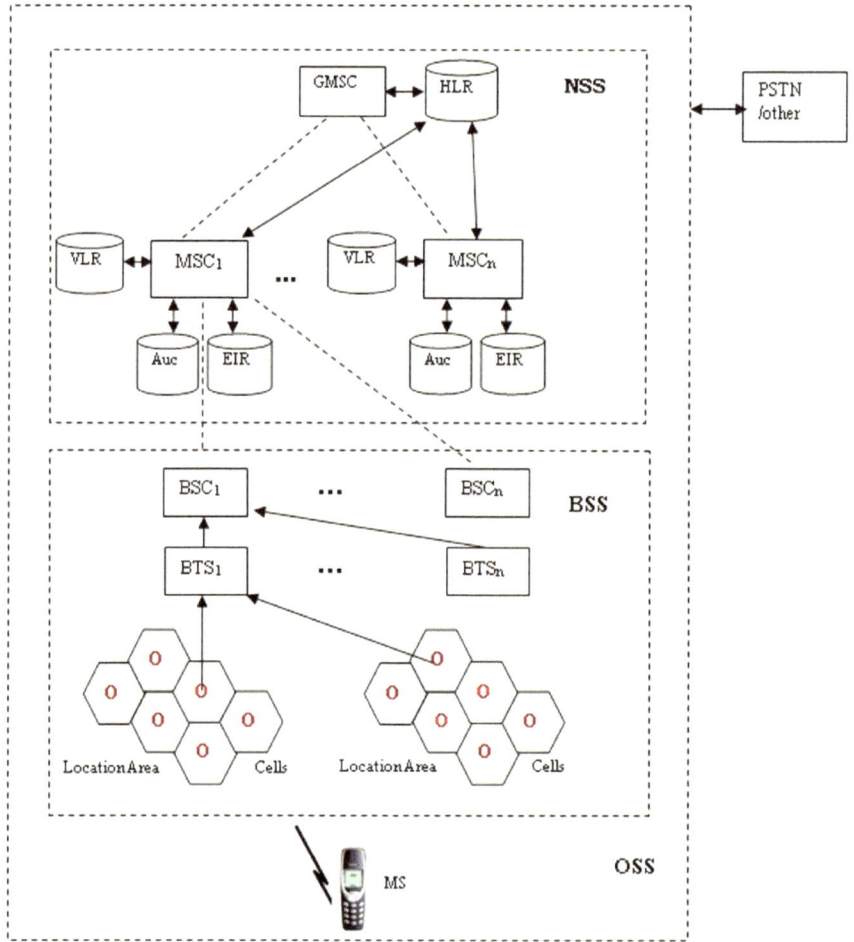

Fig. 2 GSM network architecture

Each SIM card is also uniquely identified with its International Mobile Subscriber Identity (IMSI), and it can be made to the existence of different mobile stations, but not simultaneously. This provides mobility for a mobile subscriber called personal mobility. In addition, the service provider assigns to each user a unique number known as Mobile Subscriber Integrated System Data Network (MSISDN) which corresponds to their respective SIM.

A GSM network comprises the fixed infrastructure and mobile stations. Mobile stations use the fixed infrastructure services and communicate through the radio interface. The fixed infrastructure of GSM network is divided into three subsystems, as presented in [8]. The three subsystems are present; they are:

(1) The base station subsystem (BSS)
(2) Network switching subsystem (NSS), and
(3) Operation support subsystem (OSS).

A brief description of subsystems is as follows.

1.4.1 Base Station Subsystem (BSS)

The base station subsystem (BSS) comprises GSM network components, namely the base transceiver stations (BTS) and base station controllers (BSC) which facilitate the transmission of data from a mobile subscriber to Mobile Switching Center as shown in Fig. 2.

Here, the interactions between components of a BSS are described briefly. A base transceiver station (BTS) is a system where subscribers are connected to it through radio signals in a GSM network. Location area (LA) is a set of cells in the jurisdiction of a BTS. A set of BTSs are connected to a base station controller (BSC) through the data transmission cable as shown in Fig. 3.

Fig. 3 BSS system architecture

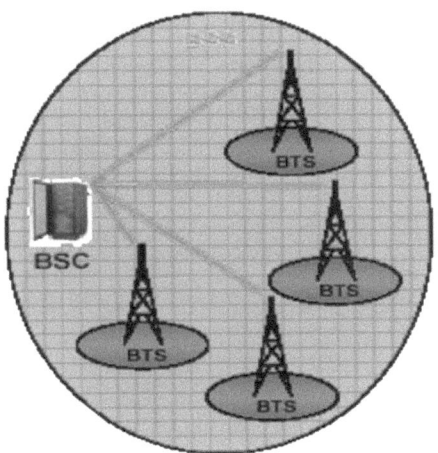

1.4.2 Network Switching Subsystem (NSS)

The network switching subsystem (NSS) of the GSM network comprises the components, namely the Mobile Switching Center (MSC) and the Gateway Mobile Switching Center (GMSC). The GMSC maintains a centralized database called the home location register (HLR) to store the files that represent the profiles of all mobile subscribers registered.

A mobile subscriber located under the jurisdiction of an MSC referred to as source MSISDN requests the MSC for a call setup giving the destination MSISDN. The record of a source MSISDN should be available at MSC for facilitating call setup. The MSC transfers the record from the HLR, and the same is stored at it in a database referred to as visitor location register (VLR). Two other databases of GSM are Authentication Center (AuC) and Equipment Identity Register (EIR). These components and databases are briefly described in the following sections.

Mobile Switching Center (MSC)

The central component of the NSS is the MSC. It acts like a normal switching node of the PSTN or ISDN and also provides all the features needed to manage a mobile subscriber, such as registration, authentication, location update, handovers, and call routing to roaming subscriber. To do this, it accesses the HLR for downloading necessary data and maintains the VLR, EIR, and AuC. The MSCs interact with a dedicated Gateway Mobile Switching Center (GMSC) for handling calls of mobile subscribers.

Home Location Register (HLR)

The persistent service profile and current location of each registered mobile subscriber are stored in HLR. When subscriber initiates a call request, HLR is first queried to determine the subscriber's current location.

The primary data stored in an HLR includes:

- IMEI: International Mobile Equipment Identity
- IMSI: MCC + MNC + MSIN: International mobile subscriber identity

    - MCC: Mobile country code
    - MNC: Mobile network code
    - MSIN: Mobile station identification

- LAI: MCC + MNC+ LAC: Location area identity

    - LAC: Location area code

- MSISDN: Mobile number (in this study assumed it as subscriber)
- MSRN: Mobile Subscriber Roaming Number

- VLR address
- LA: Location area
- Subscriptions + services
- Ref Authentication (K_i)
- PLMN info
- Electronic Serial Number (ESN)
- Mobile identification number (MIN)
- System identification code (SID)

Visitor Location Register (VLR)

A visitor location register (VLR) is a database that contains information about roaming subscribers of the service area of an MSC. The main role of VLR is to minimize the number of queries that MSCs have to make to home location register (HLR), which contains permanent subscriber's data.

The other essential data stored in a VLR in addition to HLR:

- a copy of HLR data, Plus
- TMSI: Temporary mobile subscriber identity
- LAI: Location area identity
- RAI: Routing area identity
- Location update status

The information flow between MSC, HLR and VLR each time a call request received from the mobile subscriber is described in Fig. 4.

When a subscriber in the fixed network PSTN dialing MSISDN, the local exchange identifies the number as a mobile number and establishes a connection to the PSTN GMSC. Since GMSC does not know the subscriber location or state, it

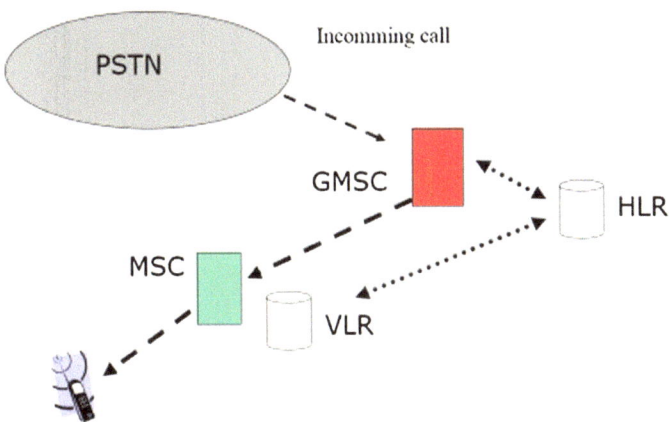

Fig. 4 Process flow between MSC, HLR, and VLR

sends a request to the HLR for information for the call that can be routed to the MSC currently serving the MS.

To route the call to the correct MSC, HLR will first request a routing number from the MSC service area's VLR. The routing number called as Mobile Subscriber Roaming Number (MSRN) contains all the necessary information to route the call request to the correct MSC. The HLR translates the MSISDN number into an IMSI before it forwards the IMSI, along with request for an MSRN, to the MSC. The MSC returns the MSRN back to the GMSC via HLR. On receiving the MSRN, the GMSC then is able to route the call directly to the correct MSC. Once the MSC is contacted, it instructs the BSCs in the subscriber location area (LA) to begin paging to MS, as LAI for the MS is known and is stored in the VLR.

A paging message is sent by the MSC to all BTS (cells) in the LA via the BSC. Once the mobile station detects its own temporary mobile subscriber identity (TMSI) to the paging channel, it responds with a paging acknowledgment message to the BTS. A traffic channel is allocated by the BSC to the MS, and the call is then setup. The mobile rings and if the mobile subscribers answer the call, connection is established.

User Authentication and Equipment Identity Register

The AuC and EIR are two other databases in NSS besides HLR and VLR. The Authentication Center (AuC) is a key component of a GSM network visitor location register (VLR). The AuC validates any security information management attempting network connection when a phone has a live network signal.

Equipment Identity Register (EIR) consults a database to determine if the service of a GSM mobile subscriber is authorized, unauthorized, or it should be monitored. It stores the serial numbers (supplied by the manufacturer) of the terminals (IMEI), which makes it possible to check for subscriber with obsolete software and to block service access for subscriber reported as stolen. Moreover, it stores information records for their subsequent processing.

1.4.3 Operation Support Subsystem (OSS)

The base station controller (BSC) and network switching system (NSS) are connected with the operations and maintenance center (OMC). The operation support subsystem is a implementation of operations and maintenance center.

The functions of OMC are:

- Administration and commercial operations
- Security management
- Network configuration, operations and performance management
- Maintenance tasks.

The network operator monitors and controls the system through its functional entity called OSS. The main objective of OSS is to facilitate the mobile user maintenance activities that are required in a GSM network such as centralized, regional, and local operations. Another significant function of OSS is to offer various operations for maintaining the network.

1.5   Models and Paradigms

In general, the required profile of a mobile subscriber is removed from VLR whenever a mobile subscriber leaves the service area of one MSC and enters the service area of another MSC. However, a phenomenon exists in the system that a mobile subscriber roams between the service area of one MSC and other several MSCs service area randomly.

Based on the accessible phenomenon, a cache model is presented in [9, 10]; model is helpful for mobile subscribers those who make and receive calls frequently and relative to the rate at which users relocate. For a record, current serving VLR is queried first for subscriber locations without contacting with HLR if it is available at VLRs cache, otherwise such required record is fetched from HLR. This approach is to find user's locations, and it is to reduce the signaling cost for receiving and delivering the calls. This model is limited, and it is performed through least replacement used (LRU) algorithm. In this model, cache entries initialization is not described in which call to mobility ratio (CMR) is increased.

In study [11], presented a local anchoring method in which transitions between VLR and HLR are reduced. Users are not required to register on every entry at VLR. Whenever, user's moves to another location from their current location, they are required to inform to nearby VLR which is referred as local anchor (LA), and later, it reports to HLR. It minimizes the cost for tracking the location whenever call arrival rate is low. If mobility rate is increased, the cost for location registration becomes high. This study discussed both static local anchor and dynamic anchor techniques. In static local anchor technique, location information of the mobile user is neither updated with HLR nor recognized. In dynamic approach, the serving VLR becomes service area for new location area (LA) for the mobile user and VLR keeps on changing their locations randomly when users are moving from one service area to another. The cost of location registration and call delivery could be minimized because records are selects dynamically from nearby VLRs.

In [12, 13], a distributed strategy model is discussed in which combined local anchors are generating replications of databases. Whenever mobile user enters one service area of a VLR, it queried for the present location of the mobile user. With this, VLRs are linked with home place and work place which are associated with a forward pointer chain. Whenever mobile user moves to a new registration area (RA), the user's new location is updated at one local anchor among multiple local anchors. Otherwise, the user's new location is informed to HLR and its replicated places. Sometimes, querying HLR is not required, if the calls initiate from one the local anchor or its replication position. It reduces the location registration cost, update cost, and cost of call delivery because all VLRs are linked with each other.

In study [14], presented a model in which a mobile user's moves forward and backward from thier neighboring service area to home service area. It reduces the unnecessary registrations at HLR for user every entry. By this, the usage of bandwidth is reduced between HLR and VLR and storage space is also reduced at VLR.

In [15], a scheme for movement-based tracking is introduced. In this scheme, cell movement counter is increased by one, whenever a mobile user visits the same cell for more than one time. The cell counter bit is set to 1 if the mobile user has visited the cell previously, or it is set to 0. This study exploits mobile user's location, movement patterns, and it predicts the paging area. Therefore, it minimizes the paging cost because it minimizes the number of searches and duplicate registrations.

The policy for least frequent replacement is discussed in [16], where the least frequently accessed mobile users are replaced in place of new mobile users. Two schemes are presented; they are inactive and random replacement. These schemes minimize the location update costs moderately. These approaches handle the overflow of mobile user's requests.

In [17], cell-based, time-based, and distance-based threshold approaches are discussed. In these approaches, a location update is done when mobile user moves away from common cells. The proposed schemes reduce the location update costs, but it is difficult to implement practically, and it needs excessive computational overhead in case of large networks.

In [18], centralized and fully distributed approaches are discussed in which HLR is to perform location update operations whenever mobile user moves across a VLR service area. The mobile user swings from VLR m to VLR n and accordingly to VLR p. The HLR and service proxy are updated at VLR m and then VLR p subsequently. In distributed approach, a mobile user shifts from VLR m to VLR n, the HLR server is updated to VLR n, and the service proxy migrates from VLR m to VLR n. Call mobility ratio is low, and both the approaches perform worse than dynamic and static approaches.

In [19], mobility management model is discussed, in which mobile users are registered in database only when mobile users make calls frequently, referred to as fVLR. This fVLR is managed mobile users data and those who are visited frequently at one VLR. In this approach, whenever user makes a call, it first queries in fVLR for required mobile user's record or data instead of requesting HLR of the called mobile user. Whenever a new user makes a call, fVLR database is updated and stored as a frequently visited user list. Mobile user data is deleted from fVLR, if no calls attempted during certain period of time.

In study [20], proposed a mechanism that defers the deletion of record of a mobile subscriber that roams randomly until the end of a block of seven days. Further, they maintain seven sets of MSISDNs one for each day. An MSISDN is included as an element of a particular set, provided it has made at least one call setup request. At the end of seventh day, the intersection of seven sets referred to as

Common Mobile Subscribers (CMS) is determined. It is carried to the first day of next block of seven days while deleting the previous block. It is proved experimentally that this mechanism minimizes the traffic load and power consumption while improving spectrum efficiency.

Hence, it is motivated to employ sliding window in place of a fixed block of seven days and evaluate its performance with regard to throughput and call setup time. Thus, the following models are presented in this book.

This book presents an algorithm that employs a sliding window of size seven, computes the intersection of seven sets of MSISDNs, slides the window right by one day that deletes the first set and makes the seventh set null and appends the intersection to the seventh set.

This book discussed a simulation model with numerical examples for evaluating the performance of the above sliding window algorithm with regard to throughput and call setup time. Optimization model is presented for determining the optimal sliding window size for which the throughput would be maximum and setup time would be minimum.

In this book a model is presented that integrates an algorithm of sliding window of size seven days with a single server finite queuing model for measurement of realistic throughput of a MSC considering the waiting times of call setup requests. It assumes that a MSC can process one call setup request at a time. A queuing theory presented for considering a MSC with multiple identical channels for processing concurrent multiple call setup requests for measuring still more realistic throughput of a MSC. Later, a simulation model considering single and multiple finite queuing channels for evaluating the performance of the above integrated models with regard to throughput and call setup time.

At the end, a cost model is presented for determining an optimal number of channels with the criterion of maximizing throughput of a MSC in which the throughput would be maximized and call setup time would be minimized.

2   Summary

Wireless mobile communications are a technology that allows transmission of data, voice, and video via a computer or any other wireless-enabled device without having to be connected to a fixed physical link. This chapter will give an overview of mobile communications, mobility, and mobility management. The Global System for Mobile communications is a European standard for digital cellular voice telecommunications, and its architecture is reviewed briefly. The importance is given to the fundamental models which is analyzed and classified with a mixture of attributes. A new approach to the problem can be developed, by classifying existing essential techniques and identifying various parameters.

Sliding Window Algorithm

1 Introduction

Mobile subscribers move randomly in the area of a GSM network. The location identity of roaming mobile subscribers is required to offer essential services to the subscriber call setup requests. In a network, subscriber information is maintained by databases, referred to as Home Location Register (HLR) and Visitor Location Register (VLR). The HLR is a centralized database which is located at Gateway Mobile Switching Center (GMSC) to maintain and keep switching profiles of all mobile subscribers and also to their current location data. VLR is distributed database in MSC to keep switching replications of subscriber profiles that are currently in its jurisdiction.

This chapter discusses an algorithm proposed in [20], a fixed block of seven days that state the keeping time for holding the subscriber profiles in VLR.

This research study is motivated from previous studies and proposes an algorithm called sliding window of size seven days in order to maximize the availability of subscriber's records in VLR. The algorithm is to determine a period of availability of records to minimize the time spent on the call setup and to maximize throughput at an MSC service area. The problem definition, method, and algorithms are presented in subsequent sections.

1.1 The Model

A mobile subscriber arrives in and departs from a MSC service area randomly. Its corresponding record must be available in the VLR for processing its call setup request. Usually, the corresponding record is retrieved from HLR to VLR on arrival and the same is removed from the VLR when it leaves. This is referred to as naïve method (NV) shown in Fig. 1.

© Springer Nature Singapore Pte Ltd. 2017
N. Mallikharjuna Rao and M. Muniratnam Naidu, *Sliding Window Algorithm for Mobile Communication Networks*, https://doi.org/10.1007/978-981-10-8473-7_2

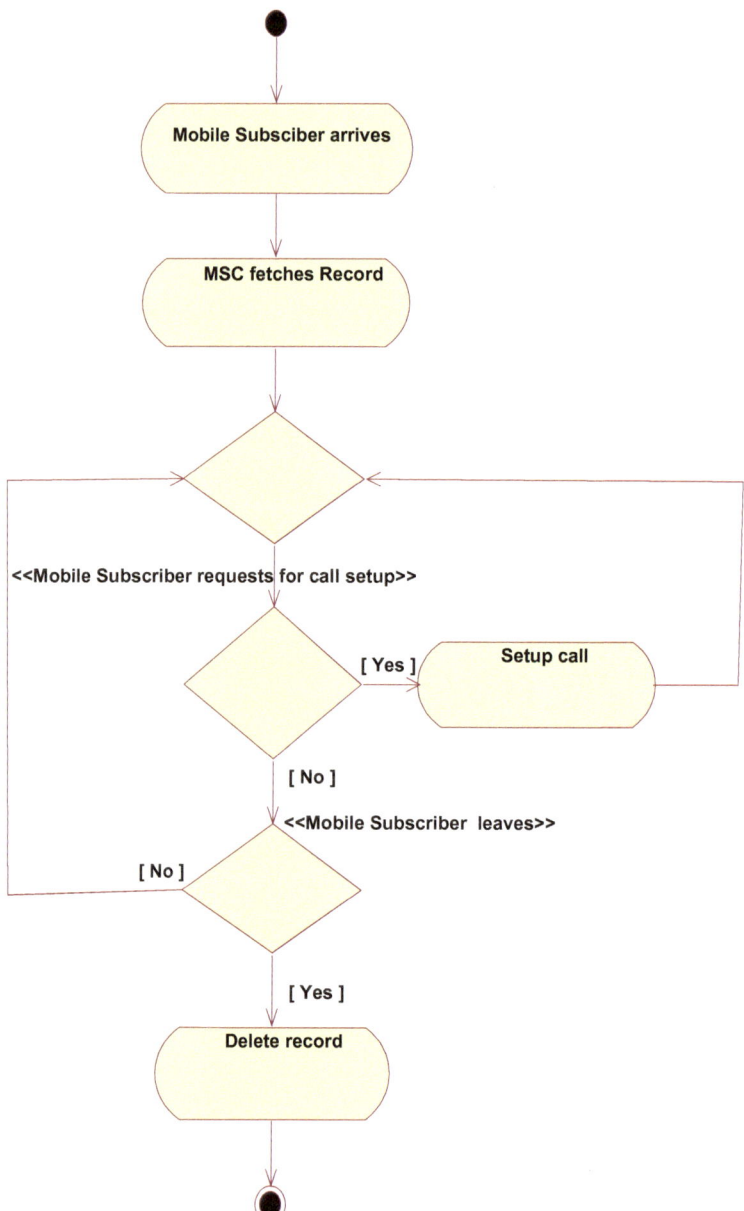

Fig. 1 Activity diagram for naïve method

The naïve approach traffic is increased on the network, which delays call setup time and minimizes the throughput of a Mobile Switching Center. Therefore, the problem is to determine the period for holding the subscriber profile (i.e., subscriber record which contains complete information about subscriber) in the VLR instead of removing the available record instantaneously when the subscriber moved to another MSC service with the aim of improving throughput of the Mobile Switching Center and reducing the call setup time. Shah et al. [24] proposed fixed block of seven days (FBSD) algorithm. It is presented in the following section.

2   Fixed Block of Seven Days (FBSD) Algorithm

2.1   *Method*

In FBSD algorithm, the relevant record of a mobile subscriber profile is fetched to the VLR from the HLR on its first arrival during a block of seven days. The MSC retains the record until the end of block for processing one or more call setup requests from the mobile subscriber. The distinct MSISDNs of the mobile subscribers who make at least one call setup request on a given day form as a set for that day. So, seven sets of MSISDN are formed over a block of seven days. After determining the intersection of seven sets, they are deleted. The complement of intersection is carried to the first day of the next block. This process is repeated during the operation of the GSM network.

Definitions of notation used to present FBSD algorithm are as follows:

$$S_i = \{s | s \text{ is an MSISDN that makes at least one call setup request on } i\text{th day}\}$$
$$\text{for } (1 \leq i \leq 7)$$

$$S_i = \{\emptyset\} \quad \text{for } (1 \leq i \leq 7)$$

$$S_i = S_i \cup s \quad \text{for } (1 \leq i \leq 7)$$

$$I = \bigcap_{i=1}^{i=7} S_i$$

$$S_1 = S_1 \cup I$$

$$R = \{r | r \text{ is a record of } s \in I\}$$

$$VLR = VLR - R'$$

The steps of FBSD algorithm are as follows:

1. $S_i \leftarrow \{\emptyset\}\ for\ (1 \le i \le 7)$

2. $i \leftarrow 1$

3. Call setup request arrival event occurs from MSISDN, s

4. If its record is found in VLR setup call

5. Otherwise

    {

    a. Fetch its record from HLR and add to VLR

    b. Setup call

    c. $S_i \leftarrow S_i \cup s$

    }

6. Repeat steps 3 through 4 until end of ith day

7. If $i < 7$

    {

    a. $i \leftarrow i + 1$

    b. Go to step 3

    }

8. Otherwise

    {

    a. Compute $I \leftarrow \cap_{i=1}^{i=7} S_i$

    b. $S_i \leftarrow \{\emptyset\}\ for\ (1 \le i \le 7)$

    c. $S_1 \leftarrow S_1 \cup I$

    d. $VLR \leftarrow VLR - R'$

    }

9. Go to step 2

The lines 1 and 2 are initialization statements. Line 3 reads the call setup requests from calling population source. Lines 4–5 execute call setup requests for the first day. Each call setup request is checked in VLR of a MSC; if such record is found, hit count increases; otherwise, miss count is increased, and such record is added to VLR. Line 6 executes call setup request for first seven days. The *if* condition that spans from 7 to 8 computes the intersection for every seven sets (days) and nullifies all the sets and carries the resultant set to the first set of next block. The complement of intersection is carried to VLR. The line 9 is executed until satisfying the stopping criterion. Finally, reports are generated with hits by weekwise for use in the performance evaluation.

Furthermore, the activity diagram shown in Fig. 2 depicts the work flow of FBSD algorithm for first day. The call setup establishment steps of FBSD algorithm are as follows:

1. A mobile subscriber arrives at an MSC service area.
2. Search relevant record at an MSC.
3. If such record is found, call setup processed.
4. Otherwise, access HLR for relevant record and process call setup.
5. Update each call setup request in set S_i in a day.
6. At the end of first day, advance by one day; otherwise, an MSC receives next call setup request till end of the first day.

The activity diagram shown in Fig. 3 depicts the work flow of FBSD algorithm for block of seven days. Start with at least one call setup request in a day, and continue till end of the day. Similarly, continue the same process till end of the seventh day. At the end of seventh day, compute the intersection of seven sets in a block and delete all the records from all the seven sets in a block and copy the complement of intersection to first set of next block and to VLR. This process is continued for every seven successive days.

2.2 Illustrative Example

The algorithm proposed by authors in [20] does not delete the record of a mobile subscriber even though it leaves the area during a block of seven days. Further, it maintains for each day independently the set of MSISDNs which request for at least one call setup. At the end of seventh day, an intersection of seven sets is determined. Seven sets are deleted, and the intersection is copied to first day. The process of determining the intersection of seven days, deleting seven sets, and copying the intersection to first day is repeated.

The algorithm considers a block of seven days. It also considers the number of call setup requests made by mobile subscribers with distinct MSISDNs which are stored in sets of seven consecutive days. It maintains the records for each day independently as set of MSISDNs which request for at least one call setup as shown

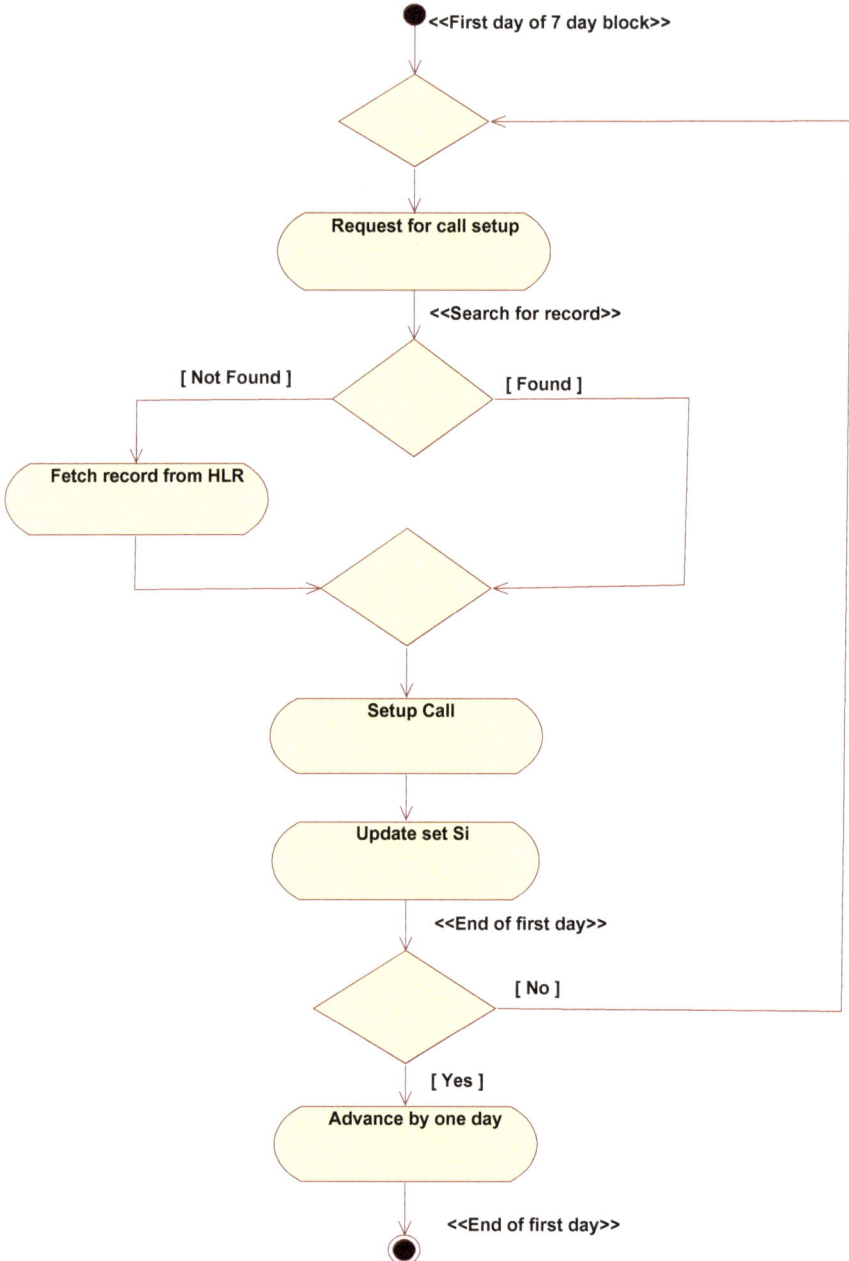

Fig. 2 Activity diagram for FBSD first-day call setup requests

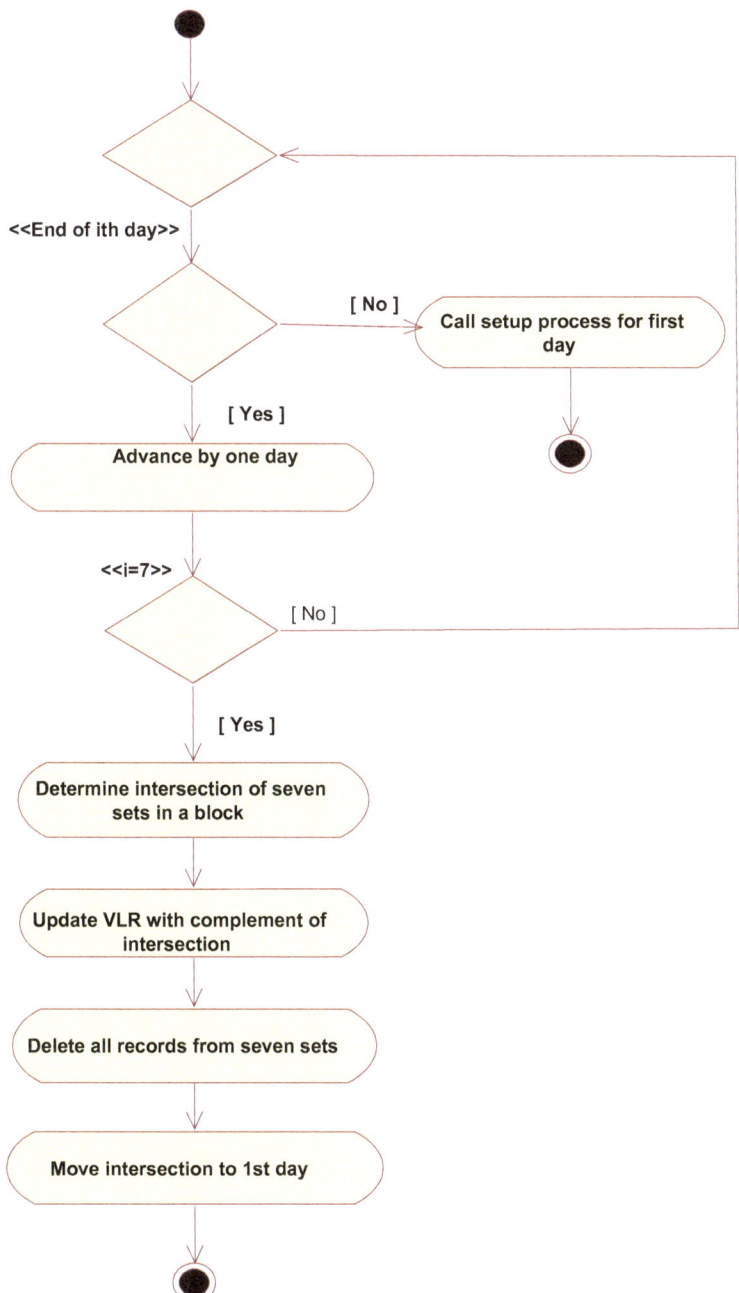

Fig. 3 Activity diagram for FBSD algorithm

(a)

Day$_1$	Day$_2$	Day$_3$	Day$_4$	Day$_5$	Day$_6$	Day$_7$
∅	∅	∅	∅	∅	∅	∅

(b)

Day$_1$	Day$_2$	Day$_3$	Day$_4$	Day$_5$	Day$_6$	Day$_7$
{MSIS-	{MSIS-	{MSIS-	{MSIS-	{MSIS-	{MSIS-	{MSIS-

(c)

Day$_1$	Day$_2$	Day$_3$	Day$_4$	Day$_5$	Day$_6$	Day$_7$
$S_i + ∅$	∅	∅	∅	∅	∅	∅

(d)

Day$_1$	Day$_2$	Day$_3$	Day$_4$	Day$_5$	Day$_6$	Day$_7$
$S_i +$	{MSIS-	{MSIS-	{MSIS-	{MSIS-	{MSIS-	{MSIS-

Fig. 4 **a** First set of seven days block with no records, **b** first set of seven days block with subscriber records, **c** after first intersection process, **d** next seven sets block with records

in Fig. 4a. The MSISDN of mobile subscriber who made at least one call setup request during ith day is calculated as per Eq. (1).

$$S_i = \{s_i/s \text{ is MSISDN of MS which made at least one call request on } i\text{th day}\}$$
$$(i = 1, 2, 3, \ldots, 7)$$

$$(1)$$

Whenever a subscriber arrives and makes a call setup request, the MSC checks in its VLR to determine whether such record is available in its VLR or not. If it is available, such record is considered as hit; otherwise, it is considered as miss. If record is missed, it is accessed from HLR and added to it VLR of an MSC. These steps are continued till end of every day using Eq. (2).

$$h = \begin{cases} 1 & \text{for } s \in S_i \\ 0 & \text{for } s \notin S_i \end{cases}$$

$$(2)$$

The similar process is to be continued for next set of days. Each time a subscriber makes a call setup request with different MSISDN, the subscriber record is kept in sets. These records are stored in sets of seven successive days. On first day, using Eq. (2), whenever a new MS arrives at the MSC service area, a query is first sent to VLR by its MSC. If such subscriber is available in VLR, a hit occurs and the record is not required to be added to VLR of MSC (i.e., store distinct MSIDSNs in VLR). If the subscriber is not available, it is treated as a miss and that record is considered as a new arrival at MSC service area. This record is added to VLR of an

MSC. This will continue for every call setup request made by subscribers on the days of sets for seven successive days as shown in Fig. 4b.

$$S_i = \bigcap_{j=1}^{7} (D_j) \quad \text{for } j = 1 \ldots 7 \tag{3}$$

where S_i = set of MSISDN records in VLR at a MSC
where D_j a call request set of jth day during ith period of seven consecutive days.

At the end of seventh day, an intersection of seven sets is determined using Eq. (3). The seven sets are deleted, and the intersection is copied to first day. Therefore, all seven sets are nullified and the complement of the intersection on the first day of seven consecutive days is copied, as shown in Fig. 4c. The process of determining the intersection of seven days, deleting seven sets, and copying the intersection to first day is repeated as shown in Fig. 4d. It is proved that this algorithm reduces the network load and power consumption and increases the efficient use of the spectrum.

Previously, FBSD algorithm is experimentally demonstrated to minimize traffic load and energy consumption ignoring overloading the database. However, in this algorithm, VLR stores all seven consecutive days' subscriber data which leads to the problem of overload in VLR, space, and time consumption. The next section discusses sliding window algorithm in detail.

3   Sliding Window of Size Seven Days Algorithm

A window can be viewed as a queue. When it slides, a number is pushed into its back, and its front is pop off. Sliding window is also known as windowing, and it is deterministic. It gives an idea about the view of memory that can be instantly shifted to another location. The number of units specified in a window is called the window size.

A sliding window is defined in [21, 22] as follows:

Definition In a given time window T, the set of records appearing in the time window $(t - T + 1, t)$ figured a glide s. Assume s_i be the ith glide assumed sliding window sw_i allied along with s_i is the set of sw successive glides from $s_i - sw + 1$ to s_i. One slot of window moves advance by one (i.e., adding the new slot $s_i - sw + 1$) with a certain amount of analysis. Thus, more records those are added to every sliding window are $|s_i|$.

3.1   Method

The sliding window of size seven days algorithm employs the model sliding window with window size seven days is proposed. First and foremost, window is maintained for each day separately the set of MSISDN which requests for at least one call setup process. At the end of seventh day, the intersection of seven sets is determined. Therefore, the sliding window is slid to right by one day in which the second set becomes first set, the third set becomes second set and so on and seventh set becomes a null set. Finally, the intersection is copied to seventh day. It is repeated to determine the intersection of seven days, slide the window right by one day, and move the intersection to seventh day.

3.2   Algorithm

The sliding window of size seven days algorithm has been proposed for determining the holding period of a record of a mobile subscriber information in VLR, rather than deleting immediately whenever mobile subscriber moves from one MSC service area to another MSC service area. The objective of proposed model is to maximize throughput of the MSC and minimize call setup time. Thus, the general concept of the proposed model is described briefly.

In first day, a call setup request from MSISDNs constitutes a first set of the sliding window. To begin with, the days are designated as numbers from 1 to 7 and their consequent sets are set to zero. Whenever a MSISDN makes a call setup request, every time it is updated in a set by applying union operation and also stored in VLR. The pertinent record of a MSISDN is fetched from HLR owing to non-availability of the same in VLR. Therefore, end of seventh day, intersection of seven sets of the sliding window is determined. The first set of sliding window is deleted by advancing remaining sets left by one position and adding new set at seventh position, and such determinant is moved to the newly added position of a sliding window. Therefore, for every step day is increased by one. Subsequently, finishing of every day, determining the intersection task is continued and resultant is copied to seventh set. This process is continuing over days. The definitions and notations are presented below:

$$S_i = \{s | s \text{ is an MSISDN that makes at least one call setup request on } i\text{th day}\}$$
$$\text{for } (1 \leq i \leq 7)$$

$S_i = \{\emptyset\}$ for$(1 \leq i \leq 7)$ $\triangleright$ MSISDNs, Initially null set

$S_i = S_i \cup s$ for$(1 \leq i \leq 7)$ $\triangleright$ updating records

$I = \bigcap\limits_{i=1}^{i=7} S_i$ $\triangleright$ computing the intersection of seven sets of the sliding window

$S_7 = S_7 \cup I$ $\triangleright$ updating seventh day

$VLR = VLR - S_1$ $\triangleright$ updating the VLR

Algorithms 1 and 2 represent the pseudocode procedure for call setup process and how the sliding window will be slide by right one position.

MSISDNs that represent call setup requests are inputs to the procedure CALLSETUP (). If line number becomes true, then call will be serviced; if not, pertinent record is fetched from HLR in order to setup or service a call setup and its update set S_i.

Algorithm 1 Call Setup Process

$CALLSETUP(MSISDN, i)$
1        $if\ (MSISDN) \ni VLR$
2          $Setup\ Call$
3    $else$
4          $Fetch\ record\ from\ HLR$
5          $Setup\ Call$
6          $S_i \leftarrow S_i \cup MSISDN$

Algorithm 2 Sliding window algorithm

$SWSSD(fp, sp)$
$\triangleright fp = day\ of\ first\ postion\ of\ window$
$\triangleright sp = day\ of\ seventh\ postion\ of\ window$

1	$I \leftarrow S_{fp}$
2	$for\ i = fp + 1\ to\ sp\ do$
3	$I \leftarrow I \bigcap S_i$
4	$sp \leftarrow sp + 1$
5	$S_{sp} \leftarrow I$
6	$fp \leftarrow fp + 1$

The statements in lines 1–3 determine the intersection of seven sets over the days from fp to sp. The line 4 is for incrementing the day of seventh position by one. The line 5 is for copying intersection into seventh position. The line 6 is for incrementing the day of first position by one.

For evaluating performance of sliding window of size seven days algorithm, a simulation model is developed. The definition of notation and the pseudocode for simulating over a given number of days is as follows:

SWSSD (λ, a, b, T)

▷ λ= Parameter of Poison Distribution

▷ a =1 and b =88243 are parameters of discrete uniform distribution

▷ T= operating life of system in days

1. $S_i \leftarrow \{\emptyset\} \ for \ (1 \leq i \leq 7)$

         ▷ set of MSISDNs for i^{th} day, initially null set

2. $h \leftarrow 0$ ▷number of hits, initially zero

3. $i \leftarrow 1$

4. $j \leftarrow 1$

5. $for \ i = 1 \ to \ 7 \ do$

    {

▷CR = Poison random variates that represents number of call setup requests for i^{th} day

6. $CR \leftarrow Generate\_Poison\_Random\_Variant(\lambda)$

7. $for \ k = 1 \ to \ CR$

    {

    ▷ Generate a discrete uniform random variant 's' that represents an MSISDN

8. $s \leftarrow Generate\_Discrete\_Uniform\_Random\_variante(a, b)$

9.   If profile of 's' is found in VLR

    {

        Setup call

        $h \leftarrow h + 1$

    }

10.   Otherwise

            {

                    a. Fetch the profile of 's' to VLR from HLR

                    b. Setup call

                    c. $S_i \leftarrow S_i \cup s$

              }

            }

          }

11.   Compute $I \leftarrow \cap_{i=1}^{7} S_i$ ▷ Intersection of seven days

12.   Slide the window right by one day

13. $VLR \leftarrow VLR - S_j$ ▷ VLR is updated

14. $j \leftarrow j + 1$

15. $S_i \leftarrow I$

16. $While(i \leq T)do$

          {

17. $CR \leftarrow Generate\_Poison\_Random\_Variant(\lambda)$

18. $for\ k = 1\ to\ CR$

            {

19. $s \leftarrow Generate\_Discrete\_Uniform\_Random\_variante(a, b)$

20.   If the profile of 's' is found in VLR

                  {

                      Setup call

                      $h \leftarrow h + 1$

                  }

21.　　Otherwise

　　　　{

　　　　　　　a.   Fetch the profile of 's' to VLR from HLR

　　　　　　　b.   Setup call

　　　　　　　c. $S_i \leftarrow S_i \cup s$

　　　　}

　　}

22. Compute $I \leftarrow \cap_{k=j}^{i} S_k$

　　　　▷ Intersection of seven days from j^{th} day

23.　　　$i \leftarrow i + 1$

　　}

The metrics employed for evaluating the performance of the SWSSD algorithms are hit rate and throughput that shall be maximized. The hit rate is the probability of the relevant record availability in VLR for processing a call setup request. The number of hits and call setup requests over a period of time is measured for computing hit rate. It is assumed that call setup requests are received randomly from a finite set of MSISDNs. The availability of relevant record of an MSISDN in VLR for processing its call setup request is referred to as a hit, whereas non-availability is referred to as a miss.

3.3   Algorithm Design Steps

In SWSSD algorithm, the call setup procedure for the first day is similar to as FBSD algorithm, which is referred in Fig. 5, activity diagram for call setup requests. Further, the activity diagram shown in Fig. 6 depicts the work flow of SWSSD algorithm.

1. An MSC receives call setup request till the end of nth day.
2. At the end of nth day, determine the intersection of first 'n' sets in a window.
3. Window shifted to right by one day such as sets (set$_2$ becomes set$_1$, set$_3$ becomes set$_2$ and so on set$_7$ becomes set$_6$).
4. Copy intersection to nth day.

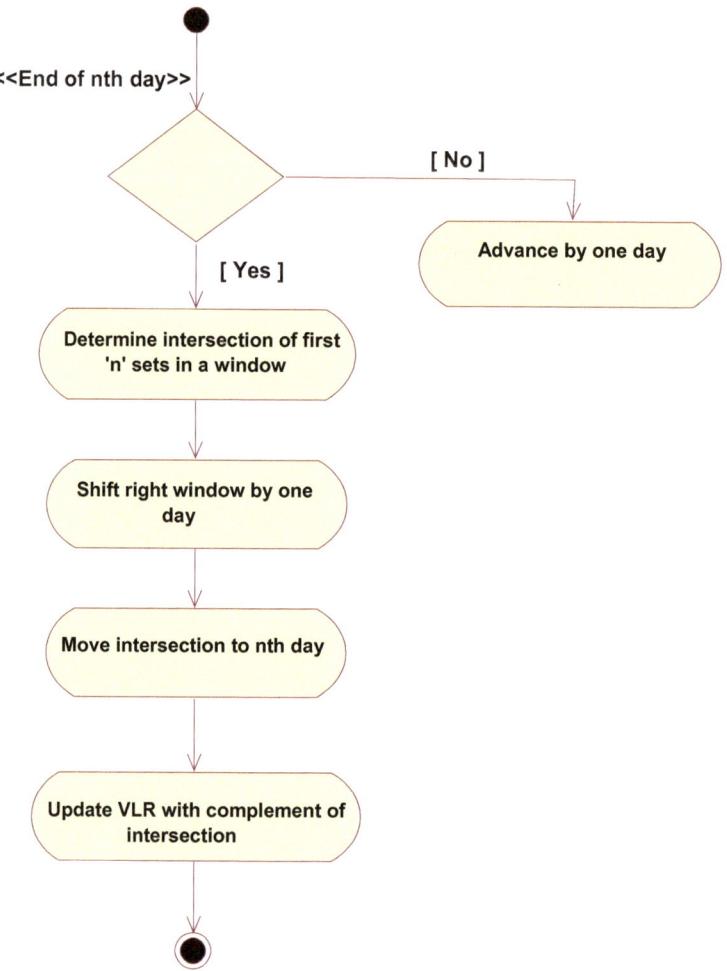

Fig. 5 Activity diagram for first '*n*' days

5. Update VLR with complement of intersection. Subsequently, in SWSSD algorithm, work flow is shown in Fig. 3.6 for end of each day.
6. At the end of *n*th day, determine intersection of seven sets in window.
7. Window shifts to right by one day.
8. Copy intersection to nth day.
9. Update VLR with complement of intersection each day.

Therefore, at the end of the time period, it determines the intersection on sliding window of size '*n*' and moves intersection to *n*th day. The records deleted from VLR except records pertaining to the determinant of intersection. The sliding

Fig. 6 Activity diagram for subsequent days

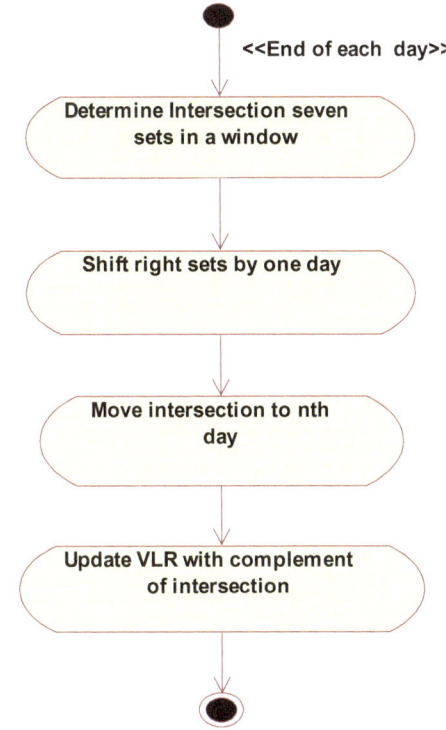

window slides to right by one day that takes the first day off and opens the eighth day. The resultant of an intersection is carried to eighth day, consequently, at the end of each day; VLR is updated with frequently requested MSISDNs.

3.4 Illustrative Example

The sliding window of size seven is proposed. Initially, it maintains for each day independently the set of MSISDNs which request for at least one call setup. At the end of seventh day, the intersection of seven sets is determined. The window slides to right by one day wherein the second set becomes first set, the third set becomes second set and so on and seventh set becomes a null set. The intersection is copied to seventh day.

The process of determining the intersection of seven days, sliding the window right by one, and copying the intersection to seventh day is repeated. The algorithm considers a window of size seven days. It considers the numbers of call setup requests made by mobile subscribers with distinct MSISDNs which are stored in set of seven days. It maintains the set of MSISDNs which request for at least one call setup as shown in Fig. 7a.

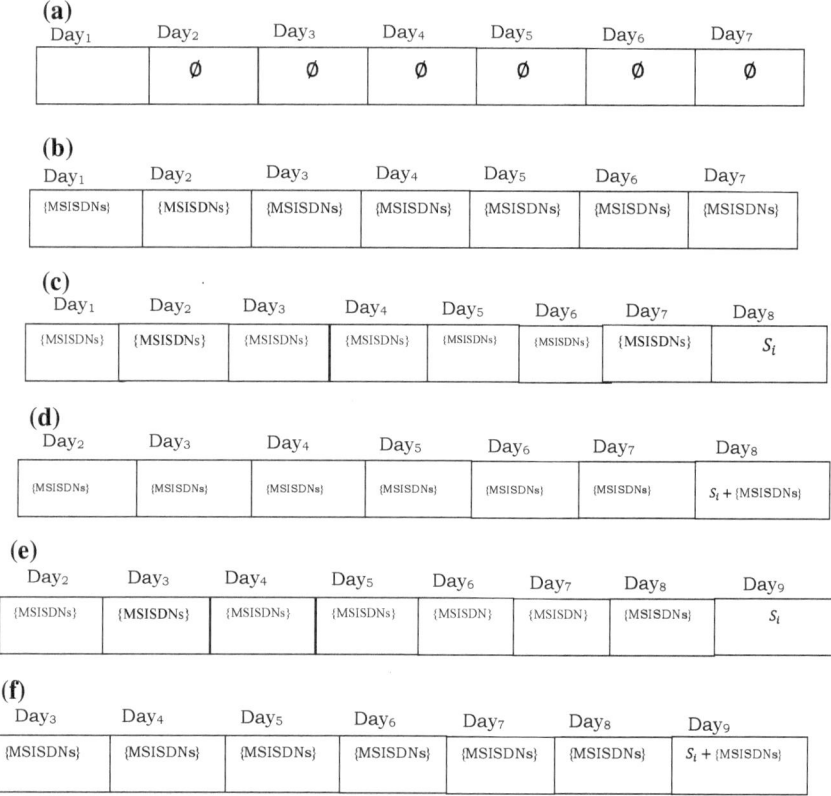

Fig. 7 a First window with seven days with day 1 records, **b** first sliding window with seven days' records, **c** after first slides of window, **d** after sliding window slide right by one day, **e** window slide right by another day, **f** after sliding window slide right by one day

The MSISDN of mobile subscriber made at least one call setup request during ith day using Eq. (4).

$$S_i = \{s/s \text{ is MSISDN of MS which made at least one call request on } i\text{th day}\}$$
$$(i = 1, 2, 3, \ldots, 7)$$

$$(4)$$

On the ith day, a mobile subscriber (i.e., MSISDN) makes a call setup request that is denoted as 's' is a record at a MSC service area. Similarly, all the requests made by subscribers in the first seven days are stored in seven sets.

Whenever a subscriber arrives and makes a call setup request, the MSC checks its database to determine whether the record is available at its VLR. If it is available, treat this as hit; otherwise, treat as miss, and such arrived subscriber record is registered with the MSC by fetching its relevant record from the HLR and added to it VLR of the MSC. It will carry on till end of the every day using Eqs. (5) and (6).

$$S_i = \begin{cases} S_i & \text{for } s \in S_i \\ S_i \cup s & \text{for } s \notin S_i \end{cases} \tag{5}$$

$$h = \begin{cases} 1 & \text{for } s \in S_i \\ 0 & \text{for } s \notin S_i \end{cases} \tag{6}$$

The similar process is to be continued for next set of seven days in a window. Whenever a new subscriber arrives at the MSC service area, a query is first sent to VLR by its MSC on a particular day, by using Eq. (6). If such subscriber is available in VLR, a hit occurs and that record does not require adding to its VLR of an MSC. If the subscriber is not available, it is treated as a miss and that record is considered as a new arrival at the MSC service area. This record is added to its VLR of an MSC. It will carry on for every call setup request made on seven consecutive days as shown in Fig. 7b.

At the end of the seventh day, the intersection of seven sets is determined using Eq. (7). The window slides to right by one day wherein the second set becomes first set, the third set becomes second set and so on and the seventh set becomes a null set. The intersection is copied to seventh day is shown in Fig. 7c.

$$S_i = \begin{cases} \emptyset & \text{for } 1 \leq i \leq 7 \\ \bigcap_{j=i-7}^{i-1} D_j & \text{where } i = 8, 9, 10 \ldots \end{cases} \tag{7}$$

S_i = set of MSISDN records in VLR at beginning of ith day.

After window slides successfully, first day set is disposed of and second set becomes first set, the third set becomes second set and new set becomes seventh set as shown in Fig. 7d. On seventh day, arrivals of call setup requests are added to complement of intersection on seventh day of next window.

At the end of next window of the seventh day, the intersection of seven sets is determined using Eq. (7). The window slides to right by one day wherein the second set becomes the first set, the third set becomes the second set and the seventh set becomes a null set. The intersection is copied to seventh day as shown in Fig. 7e. After next window slides successfully, first day is disposed of and second set becomes first set, the third set becomes second set and so on and a new set becomes the seventh set as shown in Fig. 7f. The process of determining the intersection of seven days, sliding the window right by one, and copying the intersection to seventh day is repeated.

Therefore, in SWSSD algorithm, first seven days VLR grows similar to FBSD algorithm. From second set onwards, determining intersection of sets for every day, VLR size is increased from beginning of the day and decreased at the end of each day. Hence, duplicate records are deleted through this process, and VLR size is minimized every day.

In view of the above, this research study maximizes the availability of the records in the service area of an MSC in order to minimize call setup time. If records are available at an MSC service area, call setup time minimizes; if it is not available, it will take more time to set up a call between source and destination subscribers

4   Summary

In this chapter, a model is presented with an objective of improving throughput of the MSC and decreasing the call setup time. The sliding window of size seven days algorithm has been proposed. The proposed model is to determining the record keeping time of a mobile subscriber in VLR instead of deleting it immediately.

Based on this method, an algorithm referred to as sliding window of size seven days (SWSSD) is developed. The existing algorithm referred to as fixed block of seven days (FBSD) is reviewed in detail. In next chapter, a simulation model is presented for evaluating the performance of the fixed block of seven days and sliding window of size seven days algorithms.

Performance Measurement of Sliding Window Algorithm

1 Introduction

In this chapter, a simulation model is developed for evaluating the performance of FBSD and SWSSD algorithms. Performance metrics, simulation model, input parameters, data assumptions, experimentation, and simulation output analyses are presented in the following sections.

2 Performance Metrics

The metrics employed for evaluating the performance of the FBSD and SWSSD algorithms are hit rate and throughput that shall be maximized. The performance metrics are defined below.

2.1 Hit Rate

The hit rate is the probability of the relevant record availability in VLR for processing a call setup request, and the same is computed using Eq. (1).

$$\text{Hit Rate} = \frac{\text{No. of Hits}}{\text{No. of subscriber requests}} \tag{1}$$

The number of hits and call setup requests over a period of time are measured for computing hit rate. It is assumed that call setup requests are received randomly from a finite set of MSISDNs. The availability of relevant record of an MSISDN in VLR

© Springer Nature Singapore Pte Ltd. 2017
N. Mallikharjuna Rao and M. Muniratnam Naidu, *Sliding Window Algorithm for Mobile Communication Networks*, https://doi.org/10.1007/978-981-10-8473-7_3

for processing its call setup request is referred to as a hit whereas non-availability is referred to as a miss.

2.2 Throughput

Throughput of an MSC is the number of call setups per unit time. The average call setup time for a call setup request is computed using Eq. (2).

$$\text{Average Call Setup Time (ACST)} = \frac{C_1 H + C_2 M}{H + M} \tag{2}$$

where

C_1 Call setup time in case of hit
C_2 Call setup time in case of miss
H Number of hits over a period of time
M Number of misses over a period of time.

Obviously, C_2 is greater than C_1 as the relevant record is to be fetched from HLR in case of a miss. From the analysis of data pertaining to call setup times of Telecom Regulatory Authority of India (TRAI); C_1 and C_2 are assigned 3 and 7.5 s, respectively.

The reciprocal of average call setup time is throughput, and the same is computed using the Eq. (3).

$$\text{Throughput (TP)} = \frac{1}{\text{Average Call Setup Time (ACST)}} \tag{3}$$

3 Simulation Model

The entities considered for developing simulation model to evaluate the performance of two algorithms are mobile stations and mobile switching center. Each mobile station is represented uniquely by its MSISDN. A set of mobile stations registered with a service provider constitute the calling population source from which call setup requests are received. The calling population is stored on disk as a relation of single attribute, MSISDN. In real-life GSM network, the profiles of calling population consisting of several attributes are stored in HLR at GMSC. An MSC is represented by its VLR in which the MSISDN of a mobile station is inserted on its first arrival into the service area of the MSC. The deletion of an MSISDN depends on the algorithm employed. A set of distinct MSISDNs which make at least one call setup request during a given day are stored in

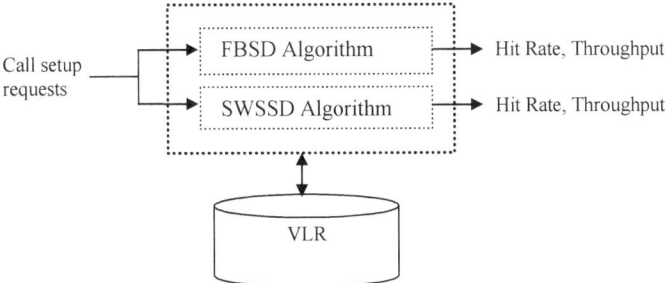

Fig. 1 Simulation model

one-dimensional array of variable size. Seven one-dimensional arrays of variable size are employed one for each day.

The call setup request arrivals per day are Poisson random variates. The MSISDNs that represent the call setup requests are discrete uniform random variates. A block diagram of the simulation model is shown in Fig. 1.

At the start of simulation period, zero is assigned to the number of call setup requests and the same is incremented by one whenever a call setup request event occurs. Similarly, zero is assigned to number of hits and the same is incremented by one whenever the relevant record of a call setup request is found in VLR.

An MSISDN, s, represents a randomly generated call setup request whereas S_i is a set of MSISDNs that make at least one call setup request on ith day which is defined in Eq. (4)

$$S_i = \{s|\ s \text{ is an MSISDN that makes atleast one call setup request on } i\text{th day}\}$$
$$\text{for}\quad (1 \leq i \leq 7) \tag{4}$$

Initially, $S_i \forall i$ are assigned null using Eq. (5)

$$S_i = \{\emptyset\}\quad \text{for } (1 \leq i \leq 7) \tag{5}$$

The VLR is searched for s. If it is not found in VLR, the same is inserted in it. Further, S_i is updated using Eq. (6)

$$S_i = S_i \cup s\quad \text{for } (1 \leq i \leq 7) \tag{6}$$

At end of the seventh day, the intersection of all seven sets is determined using the Eq. (7).

$$I = \bigcap_{i=1}^{i=7} S_i \tag{7}$$

In case of FBSD algorithm, the VLR is shrunk by retaining 'I' deleting the other records that results VLR as shown in Eq. (8)

$$VLR = I \tag{8}$$

Using Eqs. (9) and (10), S_i are updated.

$$S_1 = S_1 \cup I \tag{9}$$

$$S_i = \{\emptyset\} \quad \text{for } (2 \leq i \leq 7) \tag{10}$$

The above process is repeated until the end of simulation period. The hit rate and throughput are computed using Eqs. (1) and (3).

In case of SWSSD algorithm, the VLR is shrunk by deleting the S_1 records that result VLR as shown in Eq. (11)

$$VLR = VLR - S_1 \tag{11}$$

Using the Eqs. (12) and (13), S_i are updated.

$$S_{i-1} = S_i \quad \text{for } 2 \leq i \leq 7 \tag{12}$$

$$S_7 = I. \tag{13}$$

The above process of updating using Eqs. (12) and (13) is equivalent to sliding the window right by one day.

This process is continued till the end of simulation period. The hit rate and throughput are computed using Eqs. (1) and (3).

3.1 Simulation Parameters

The input parameters of simulation are shown in Table 1.

The sample Poisson random variates that represent call setup requests per day are shown in Fig. 2. Further, the sample discrete uniform random variates that represent call setup requests are shown in Fig. 3.

Table 1 Simulation input parameters

Parameter	Description	Value
N	Simulation run length in days	1001
M	Calling population size	77,260
M	Ordered set of calling population	$\{s \mid s \text{ is}(i, \text{MSISDN})\}$ for $1 \leq i \leq 77,260$
λ	Parameter of Poisson distribution, i.e., mean number of call setup requests per day	87

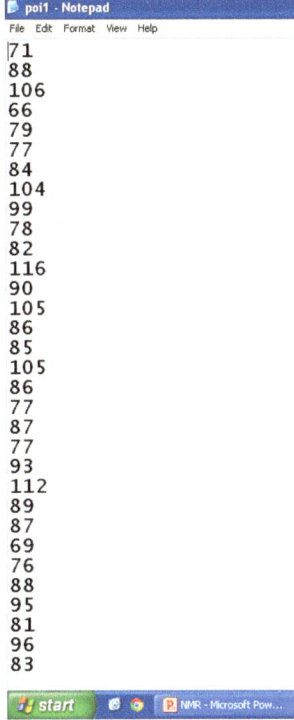

Fig. 2 Sample Poisson random variates

Fig. 3 Sample discrete random variates for a given day

4 Experimentation

The simulation is performed using TN3270 emulator for IBM (zSeries) mainframe on Windows platform. The sample screenshots are shown in Figs. 4, 5, 6, 7, 8, 9, and 10.

5 Simulation Output Analysis

The call setup requests and the uniform random variates are generated for each day over a period of 1001 days. The call setup requests for each block of seven days are aggregated. The number of hits is determined for each of 143 blocks when the call setup requests are processed employing FBSD and SWSSD algorithms. The input and output of simulation for each block are given in Table 2.

The performance metrics, hit rate, average call setup time, and throughput, are computed for each block, and the same are given in Table 3; the corresponding graphical representations are shown in Figs. 11 and 12.

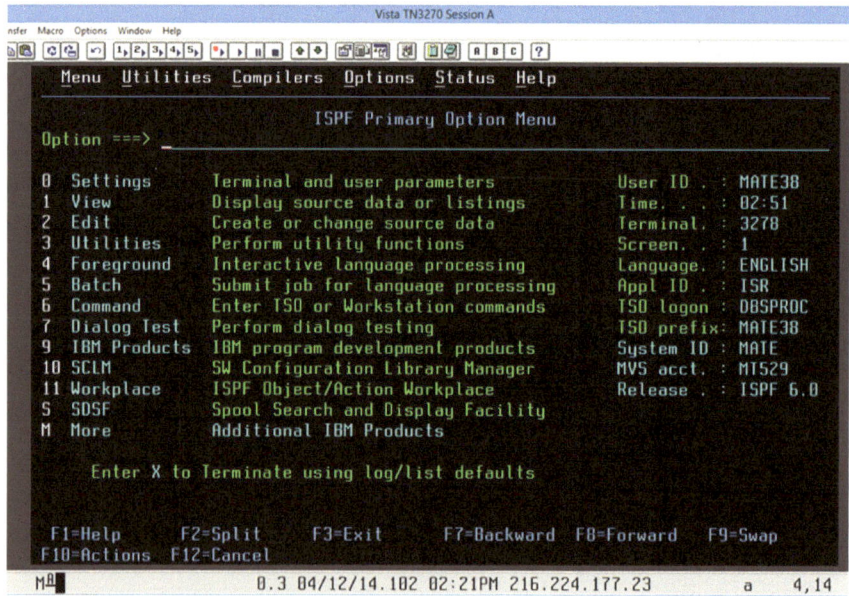

Fig. 4 ISPF 6.0 editor

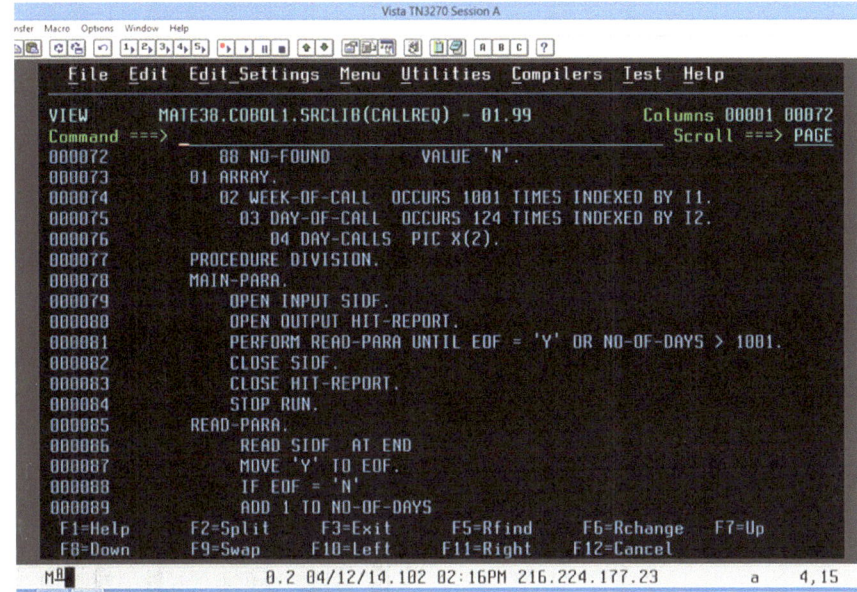

Fig. 5 IBM programming interface

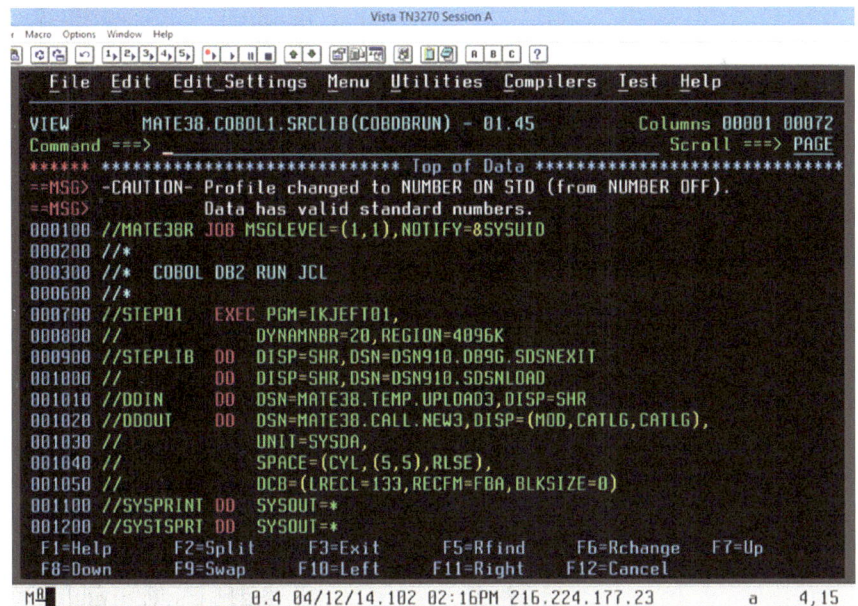

Fig. 6 JCL environment

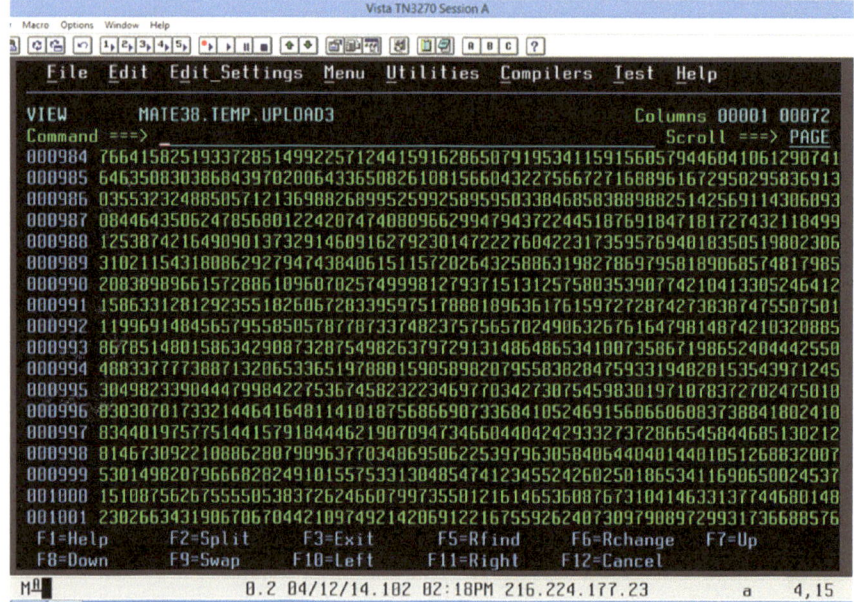

Fig. 7 Input data view

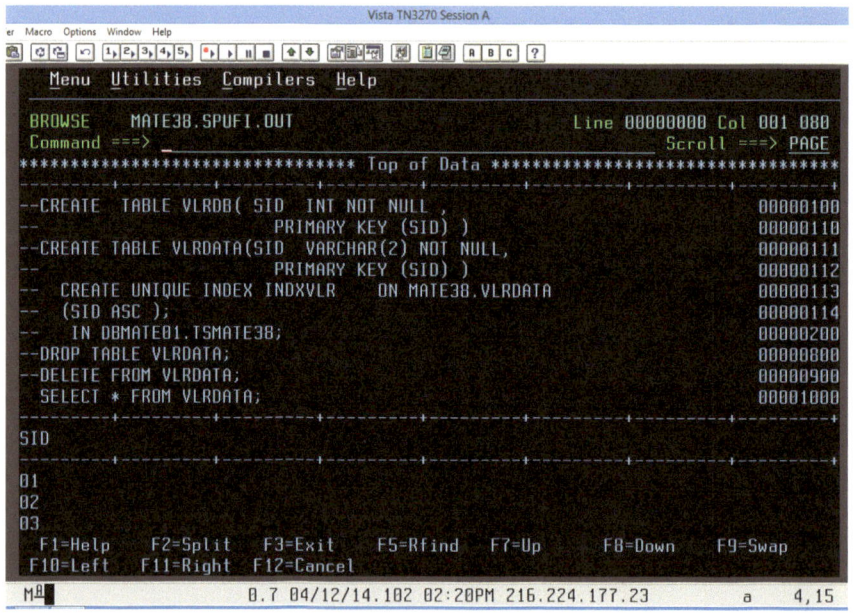

Fig. 8 DB2 database environment IBM technologies

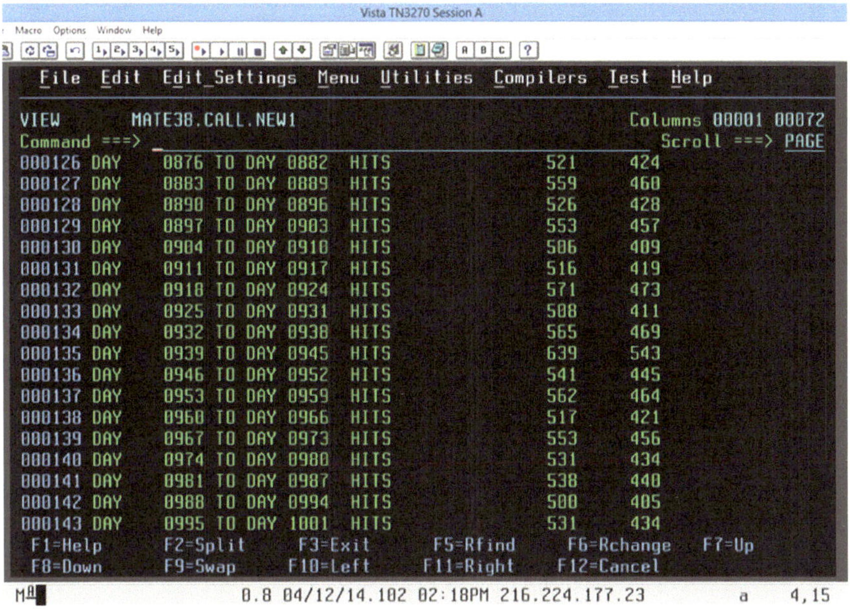

Fig. 9 System-generated output for FBSD algorithm

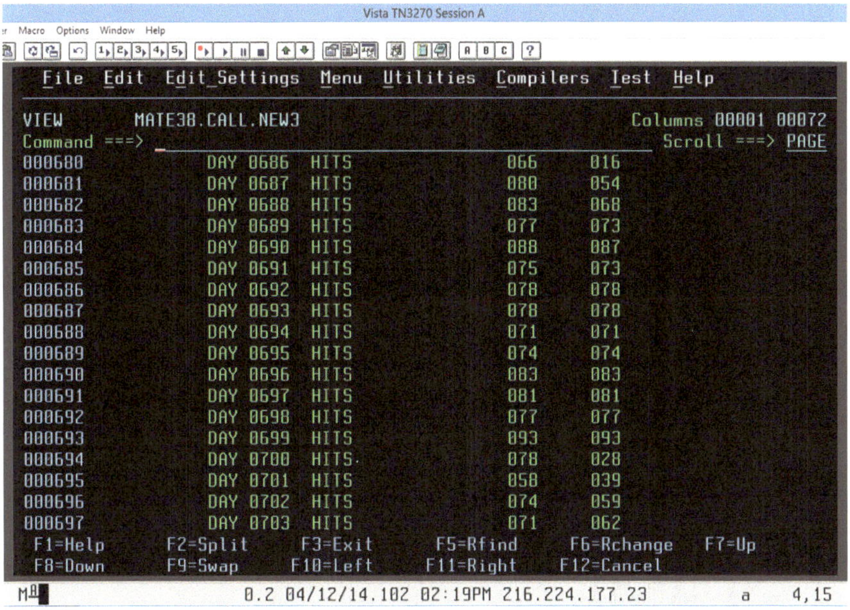

Fig. 10 System-generated output for SWSSD algorithm

Table 2 Input and output of simulation for each block

Block No.	Period		Input	Output			
	From	To	Call requests	FBSD		SWSSD	
				Hits	Misses	Hits	Hits
1	1	7	522	423	99	423	99
2	8	14	548	450	98	488	60
3	15	21	546	449	97	424	122
4	22	28	526	430	96	481	45
5	29	35	533	435	98	425	108
6	36	42	544	445	99	481	63
7	43	49	534	437	97	426	108
8	50	56	513	416	97	456	57
9	57	63	557	461	96	427	130
10	64	70	554	459	95	508	46
11	71	77	523	425	98	428	95
12	78	84	540	444	96	492	48
13	85	91	568	472	96	429	139
14	92	98	522	425	97	468	54
15	99	105	519	421	98	430	89
16	106	112	506	411	95	454	52
17	113	119	558	461	97	431	127
18	120	126	533	437	96	479	54
19	127	133	499	400	99	432	67
20	134	140	601	504	97	548	53
21	141	147	513	415	98	433	80
22	148	154	510	415	95	465	45
23	155	161	539	441	98	434	105
24	162	168	517	419	98	470	47
25	169	175	552	454	98	435	117
26	176	182	567	469	98	514	53
27	183	189	508	414	94	436	72
28	190	196	512	415	97	460	52
29	197	203	505	407	98	437	68
30	204	210	544	448	96	483	61
31	211	217	526	429	97	438	88
32	218	224	521	424	97	467	54
33	225	231	521	425	96	439	82
34	232	238	504	407	97	453	51
35	239	245	510	412	98	440	70
36	246	252	556	458	98	508	48
37	253	259	537	439	98	441	96
38	260	266	527	433	94	468	59
39	267	273	545	446	99	442	103
40	274	280	531	432	99	472	59
41	281	287	566	469	97	443	123
42	288	294	559	463	96	506	53
43	295	301	566	469	97	444	122

(continued)

Table 2 (continued)

Block No.	Period		Input	Output			
	From	To	Call requests	FBSD		SWSSD	
				Hits	Misses	Hits	Hits
44	302	308	523	425	98	481	42
45	309	315	555	457	98	445	110
46	316	322	545	447	98	494	51
47	323	329	558	462	96	446	112
48	330	336	570	472	98	519	51
49	337	343	497	400	97	447	50
50	344	350	533	436	97	479	54
51	351	357	538	441	97	448	90
52	358	364	564	465	99	508	56
53	365	371	548	451	97	449	99
54	372	378	567	471	96	514	53
55	379	385	515	417	98	450	65
56	386	392	539	443	96	491	48
57	393	399	548	452	96	451	97
58	400	406	531	432	99	473	58
59	407	413	575	478	97	452	123
60	414	420	545	447	98	486	59
61	421	427	571	476	95	453	118
62	428	434	562	464	98	508	54
63	435	441	548	450	98	454	94
64	442	448	497	401	96	449	48
65	449	455	544	446	98	455	89
66	456	462	516	420	96	474	42
67	463	469	581	482	99	456	125
68	470	476	536	439	97	482	54
69	477	483	552	456	96	457	95
70	484	490	558	460	98	501	57
71	491	497	535	437	98	458	77
72	498	504	553	454	99	506	47
73	505	511	537	439	98	459	78
74	512	518	523	426	97	466	57
75	519	525	508	412	96	460	48
76	526	532	527	428	99	472	55
77	533	539	562	465	97	461	101
78	540	546	541	444	97	492	49
79	547	553	518	423	95	462	56
80	554	560	542	443	99	492	50
81	561	567	514	416	98	463	51
82	568	574	540	445	95	485	55
83	575	581	537	438	99	464	73
84	582	588	527	430	97	474	53
85	589	595	534	436	98	465	69
86	596	602	574	475	99	523	51

(continued)

Table 2 (continued)

Block No.	Period		Input	Output			
	From	To	Call requests	FBSD		SWSSD	
				Hits	Misses	Hits	Hits
87	603	609	559	462	97	466	93
88	610	616	485	386	99	438	47
89	617	623	548	449	99	467	81
90	624	630	585	490	95	518	67
91	631	637	544	446	98	468	76
92	638	644	575	476	99	520	55
93	645	651	550	454	96	469	81
94	652	658	563	468	95	509	54
95	659	665	537	440	97	470	67
96	666	672	537	439	98	484	53
97	673	679	546	448	98	471	75
98	680	686	535	436	99	485	50
99	687	693	559	460	99	472	87
100	694	700	557	463	94	507	50
101	701	707	537	441	96	473	64
102	708	714	538	441	97	478	60
103	715	721	582	487	95	474	108
104	722	728	553	457	96	505	48
105	729	735	561	464	97	475	86
106	736	742	529	432	97	468	61
107	743	749	593	495	98	476	117
108	750	756	497	400	97	442	55
109	757	763	536	440	96	477	59
110	764	770	503	410	93	459	44
111	771	777	534	437	97	478	56
112	778	784	525	426	99	476	49
113	785	791	517	418	99	479	38
114	792	798	521	425	96	475	46
115	799	805	546	448	98	480	66
116	806	812	526	429	97	459	67
117	813	819	597	500	97	481	116
118	820	826	529	433	96	483	46
119	827	833	573	477	96	482	91
120	834	840	522	427	95	471	51
121	841	847	524	427	97	483	41
122	848	854	562	467	95	516	46
123	855	861	541	447	94	484	57
124	862	868	572	477	95	515	57
125	869	875	555	461	94	485	70
126	876	882	521	424	97	463	58
127	883	889	559	460	99	486	73
128	890	896	526	428	98	479	47

(continued)

Table 2 (continued)

| Block No. | Period | | Input | Output | | | |
| | From | To | Call requests | FBSD | | SWSSD | |
				Hits	Misses	Hits	Hits
129	897	903	553	457	96	487	66
130	904	910	506	409	97	469	37
131	911	917	516	419	97	488	28
132	918	924	571	473	98	512	59
133	925	931	508	411	97	489	19
134	932	938	565	469	96	512	53
135	939	945	639	543	96	490	149
136	946	952	541	445	96	491	50
137	953	959	562	464	98	491	71
138	960	966	517	421	96	465	52
139	967	973	553	456	97	492	61
140	974	980	531	434	97	481	50
141	981	987	538	440	98	493	45
142	988	994	500	405	95	457	43
143	995	1001	531	434	97	494	37
Total			77260	62950	13779	66945	9784

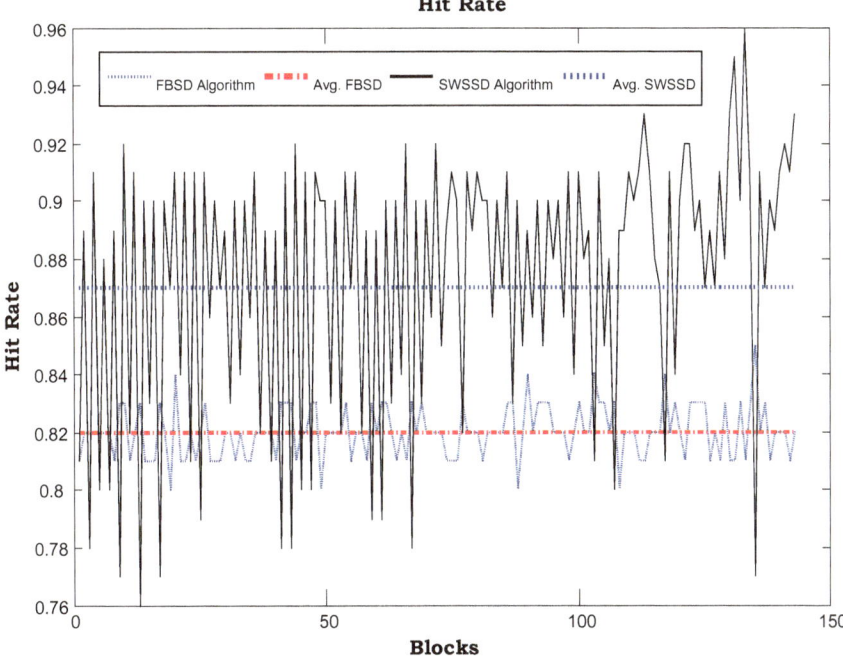

Fig. 11 Hit rate versus blocks at an MSC

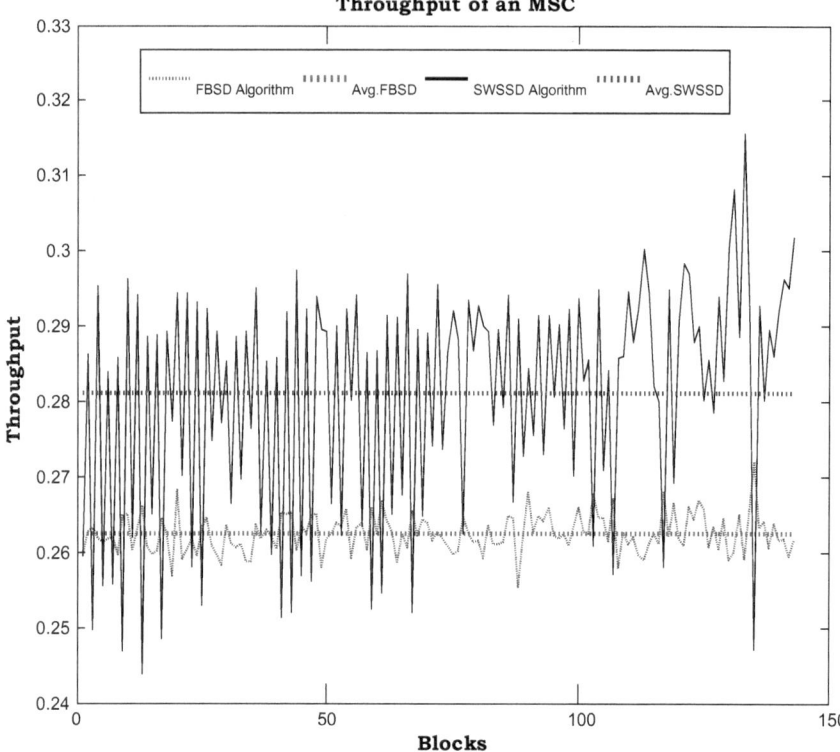

Fig. 12 Throughput versus blocks at an MSC

Table 3 Computing of performance metrics

Block No.	Performance metrics					
	FBSD algorithm			SWSSD algorithm		
	Hit rate	ACST	Throughput	Hit rate	ACST	Throughput
1	0.8103	3.8534	0.2595	0.8103	3.8534	0.2595
2	0.8212	3.8047	0.2628	0.8905	3.4927	0.2863
3	0.8223	3.7995	0.2632	0.7766	4.0055	0.2497
4	0.8175	3.8213	0.2617	0.9144	3.3850	0.2954
5	0.8161	3.8274	0.2613	0.7974	3.9118	0.2556
6	0.8180	3.8189	0.2619	0.8842	3.5211	0.2840
7	0.8184	3.8174	0.2620	0.7978	3.9101	0.2557
8	0.8109	3.8509	0.2597	0.8889	3.5000	0.2857
9	0.8276	3.7756	0.2649	0.7666	4.0503	0.2469
10	0.8285	3.7717	0.2651	0.9170	3.3736	0.2964
11	0.8126	3.8432	0.2602	0.8184	3.8174	0.2620
12	0.8222	3.8000	0.2632	0.9111	3.4000	0.2941
13	0.8310	3.7606	0.2659	0.7553	4.1012	0.2438

(continued)

Table 3 (continued)

Block No.	Performance metrics					
	FBSD algorithm			SWSSD algorithm		
	Hit rate	ACST	Throughput	Hit rate	ACST	Throughput
14	0.8142	3.8362	0.2607	0.8966	3.4655	0.2886
15	0.8112	3.8497	0.2598	0.8285	3.7717	0.2651
16	0.8123	3.8449	0.2601	0.8972	3.4625	0.2888
17	0.8262	3.7823	0.2644	0.7724	4.0242	0.2485
18	0.8199	3.8105	0.2624	0.8987	3.4559	0.2894
19	0.8016	3.8928	0.2569	0.8657	3.6042	0.2775
20	0.8386	3.7263	0.2684	0.9118	3.3968	0.2944
21	0.8090	3.8596	0.2591	0.8441	3.7018	0.2701
22	0.8137	3.8382	0.2605	0.9118	3.3971	0.2944
23	0.8182	3.8182	0.2619	0.8052	3.8766	0.2580
24	0.8104	3.853	0.2595	0.9091	3.4091	0.2933
25	0.8225	3.7989	0.2632	0.7880	3.9538	0.2529
26	0.8272	3.7778	0.2647	0.9065	3.4206	0.2923
27	0.8150	3.8327	0.2609	0.8583	3.6378	0.2749
28	0.8105	3.8525	0.2596	0.8984	3.457	0.2893
29	0.8059	3.8733	0.2582	0.8653	3.6059	0.2773
30	0.8235	3.7941	0.2636	0.8879	3.5046	0.2853
31	0.8156	3.8298	0.2611	0.8327	3.7529	0.2665
32	0.8138	3.8378	0.2606	0.8964	3.4664	0.2885
33	0.8157	3.8292	0.2612	0.8426	3.7083	0.2697
34	0.8075	3.8661	0.2587	0.8988	3.4554	0.2894
35	0.8078	3.8647	0.2588	0.8627	3.6176	0.2764
36	0.8237	3.7932	0.2636	0.9137	3.3885	0.2951
37	0.8175	3.8212	0.2617	0.8212	3.8045	0.2628
38	0.8216	3.8027	0.2630	0.8880	3.5038	0.2854
39	0.8183	3.8174	0.2620	0.8110	3.8505	0.2597
40	0.8136	3.839	0.2605	0.8889	3.5	0.2857
41	0.8286	3.7712	0.2652	0.7827	3.9779	0.2514
42	0.8283	3.7728	0.2651	0.9052	3.4267	0.2918
43	0.8286	3.7712	0.2652	0.7845	3.97	0.2519
44	0.8126	3.8432	0.2602	0.9197	3.3614	0.2975
45	0.8234	3.7946	0.2635	0.8018	3.8919	0.2569
46	0.8202	3.8092	0.2625	0.9064	3.4211	0.2923
47	0.8280	3.7742	0.2650	0.7993	3.9032	0.2562
48	0.8281	3.7737	0.2650	0.9105	3.4026	0.2939
49	0.8048	3.8783	0.2578	0.8994	3.4527	0.2896
50	0.8180	3.8189	0.2619	0.8987	3.4559	0.2894
51	0.8197	3.8113	0.2624	0.8327	3.7528	0.2665
52	0.8245	3.7899	0.2639	0.9007	3.4468	0.2901

(continued)

Table 3 (continued)

Block No.	Performance metrics					
	FBSD algorithm			SWSSD algorithm		
	Hit rate	ACST	Throughput	Hit rate	ACST	Throughput
53	0.8230	3.7965	0.2634	0.8193	3.813	0.2623
54	0.8307	3.7619	0.2658	0.9065	3.4206	0.2923
55	0.8097	3.8563	0.2593	0.8738	3.568	0.2803
56	0.8219	3.8015	0.2631	0.9109	3.4007	0.2941
57	0.8248	3.7883	0.2640	0.8230	3.7965	0.2634
58	0.8136	3.839	0.2605	0.8908	3.4915	0.2864
59	0.8313	3.7591	0.2660	0.7861	3.9626	0.2524
60	0.8202	3.8092	0.2625	0.8917	3.4872	0.2868
61	0.8336	3.7487	0.2668	0.7933	3.9299	0.2545
62	0.8256	3.7847	0.2642	0.9039	3.4324	0.2913
63	0.8212	3.8047	0.2628	0.8285	3.7719	0.2651
64	0.8068	3.8692	0.2585	0.9034	3.4346	0.2912
65	0.8199	3.8107	0.2624	0.8364	3.7362	0.2677
66	0.8140	3.8372	0.2606	0.9186	3.3663	0.2971
67	0.8296	3.7668	0.2655	0.7849	3.9682	0.2520
68	0.8190	3.8144	0.2622	0.8993	3.4534	0.2896
69	0.8261	3.7826	0.2644	0.8279	3.7745	0.2649
70	0.8244	3.7903	0.2638	0.8978	3.4597	0.2890
71	0.8168	3.8243	0.2615	0.8561	3.6477	0.2741
72	0.8210	3.8056	0.2628	0.9150	3.3825	0.2956
73	0.8175	3.8212	0.2617	0.8547	3.6536	0.2737
74	0.8145	3.8346	0.2608	0.8910	3.4904	0.2865
75	0.8110	3.8504	0.2597	0.9055	3.4252	0.2920
76	0.8121	3.8454	0.2601	0.8956	3.4696	0.2882
77	0.8274	3.7767	0.2648	0.8203	3.8087	0.2626
78	0.8207	3.8068	0.2627	0.9094	3.4076	0.2935
79	0.8166	3.8253	0.2614	0.8919	3.4865	0.2868
80	0.8173	3.822	0.2616	0.9077	3.4151	0.2928
81	0.8093	3.858	0.2592	0.9008	3.4465	0.2901
82	0.8241	3.7917	0.2637	0.8981	3.4583	0.2892
83	0.8156	3.8296	0.2611	0.8641	3.6117	0.2769
84	0.8159	3.8283	0.2612	0.8994	3.4526	0.2896
85	0.8165	3.8258	0.2614	0.8708	3.5815	0.2792
86	0.8275	3.7761	0.2648	0.9111	3.3998	0.2941
87	0.8265	3.7809	0.2645	0.8336	3.7487	0.2668
88	0.7959	3.9186	0.2552	0.9031	3.4361	0.2910
89	0.8193	3.813	0.2623	0.8522	3.6651	0.2728
90	0.8376	3.7308	0.2680	0.8855	3.5154	0.2845
91	0.8199	3.8107	0.2624	0.8603	3.6287	0.2756

(continued)

Table 3 (continued)

Block No.	Performance metrics					
	FBSD algorithm			SWSSD algorithm		
	Hit rate	ACST	Throughput	Hit rate	ACST	Throughput
92	0.8278	3.7748	0.2649	0.9043	3.4304	0.2915
93	0.8255	3.7855	0.2642	0.8527	3.6627	0.2730
94	0.8313	3.7593	0.2660	0.9041	3.4316	0.2914
95	0.8194	3.8128	0.2623	0.8752	3.5615	0.2808
96	0.8175	3.8212	0.2617	0.9013	3.4441	0.2903
97	0.8205	3.8077	0.2626	0.8626	3.6181	0.2764
98	0.8150	3.8327	0.2609	0.9065	3.4206	0.2923
99	0.8229	3.797	0.2634	0.8444	3.7004	0.2702
100	0.8312	3.7594	0.2660	0.9102	3.4039	0.2938
101	0.8212	3.8045	0.2628	0.8808	3.5363	0.2828
102	0.8197	3.8113	0.2624	0.8885	3.5019	0.2856
103	0.8368	3.7345	0.2678	0.8144	3.8351	0.2608
104	0.8264	3.7812	0.2645	0.9132	3.3906	0.2949
105	0.8271	3.7781	0.2647	0.8467	3.6898	0.2710
106	0.8166	3.8251	0.2614	0.8847	3.5189	0.2842
107	0.8347	3.7437	0.2671	0.8027	3.8879	0.2572
108	0.8048	3.8783	0.2578	0.8893	3.498	0.2859
109	0.8209	3.806	0.2627	0.8899	3.4953	0.2861
110	0.8151	3.832	0.2610	0.9125	3.3936	0.2947
111	0.8184	3.8174	0.2620	0.8951	3.4719	0.2880
112	0.8114	3.8486	0.2598	0.9067	3.42	0.2924
113	0.8085	3.8617	0.2590	0.9265	3.3308	0.3002
114	0.8157	3.8292	0.2612	0.9117	3.3973	0.2944
115	0.8205	3.8077	0.2626	0.8791	3.544	0.2822
116	0.8156	3.8298	0.2611	0.8726	3.5732	0.2799
117	0.8375	3.7312	0.2680	0.8057	3.8744	0.2581
118	0.8185	3.8166	0.2620	0.9130	3.3913	0.2949
119	0.8325	3.7539	0.2664	0.8412	3.7147	0.2692
120	0.8180	3.819	0.2619	0.9023	3.4397	0.2907
121	0.8149	3.833	0.2609	0.9218	3.3521	0.2983
122	0.8310	3.7607	0.2659	0.9181	3.3683	0.2969
123	0.8262	3.7819	0.2644	0.8946	3.4741	0.2878
124	0.8339	3.7474	0.2669	0.9003	3.4484	0.2900
125	0.8306	3.7622	0.2658	0.8739	3.5676	0.2803
126	0.8138	3.8378	0.2606	0.8887	3.501	0.2856
127	0.8229	3.797	0.2634	0.8694	3.5877	0.2787
128	0.8137	3.8384	0.2605	0.9106	3.4021	0.2939
129	0.8264	3.7812	0.2645	0.8807	3.5371	0.2827
130	0.8083	3.8626	0.2589	0.9269	3.3291	0.3004

(continued)

Table 3 (continued)

Block No.	Performance metrics					
	FBSD algorithm			SWSSD algorithm		
	Hit rate	ACST	Throughput	Hit rate	ACST	Throughput
131	0.8120	3.8459	0.2600	0.9457	3.2442	0.3082
132	0.8284	3.7723	0.2651	0.8967	3.465	0.2886
133	0.8091	3.8593	0.2591	0.9626	3.1683	0.3156
134	0.8301	3.7646	0.2656	0.9062	3.4221	0.2922
135	0.8498	3.6761	0.2720	0.7668	4.0493	0.2470
136	0.8226	3.7985	0.2633	0.9076	3.4159	0.2927
137	0.8256	3.7847	0.2642	0.8737	3.5685	0.2802
138	0.8143	3.8356	0.2607	0.8994	3.4526	0.2896
139	0.8246	3.7893	0.2639	0.8897	3.4964	0.2860
140	0.8173	3.822	0.2616	0.9058	3.4237	0.2921
141	0.8178	3.8197	0.2618	0.9164	3.3764	0.2962
142	0.8100	3.855	0.2594	0.9140	3.387	0.2952
143	0.8173	3.822	0.2616	0.9303	3.3136	0.3018
Average	0.8200	3.783	0.2625	0.8737	3.5454	0.2811

Table 4 Comparison of performance metrics

Performance metric	FBSD algorithm	SWSSD algorithm	% increase (%)
Average hit rate	0.8200	0.8737	6.54
Average call setup time	3.7830	3.5454	6.28
Average throughput	0.2625	0.2811	7.08

It is obviously evident from Table 4 that the performance of the proposed SWSSD algorithm is better than the FBSD algorithm. It is obvious that the performance of SWSSD algorithm is significantly better than that of FBSD algorithm.

6 Summary

In this chapter, a simulation model for evaluating the performance of FBSD and SWSSD algorithms is developed. It is used to generate the performance measures of FBSD and SWSSD algorithms over a period of 1001 days. It is evident from Table 4; there is a significant increase in performance metrics of the proposed

SWSSD algorithm for one MSC service area. Obviously, it is recommended to consider adopting the SWSSD algorithm for the entire GSM network for improving its throughput by 7.08%.

The next chapter proposes a model to determine the optimal sliding window size based on performance metric, throughput.

A Model for Determining Optimal Sliding Window Size

1 Introduction

In this chapter, a model is proposed to determine the optimal sliding window size (OSWS) minimizing the average call setup time that is equivalent to maximizing the throughput.

FBSD algorithm proposed in [20] suggests a seven-day holding period for the record of a subscriber in VLR. Nuka and Naidu [23] proposed an SWSSD algorithm which suggests a sliding window of size seven days for determining holding period of the record of a subscriber in VLR considering its stochastic nature of arrivals and departures that strikes trade-off between network overhead and call setup time. Its performance proves to be better than that of the FBSD algorithm.

As a further study, it is attempted to investigate the behavior of average call setup time as a function of sliding window size through simulation. It is found that the average call setup time initially decreases and subsequently increases as the sliding window size increases. Hence, it is concluded that there exists an optimal sliding window size. The model is presented in the following section.

2 The Model

The notation and assumptions of the proposed model are given hereunder.

Notation:

ws	Sliding window size in days
T	Simulation period in days
R	Number of call setup requests during simulation period
H	$g(\text{ws})$, number of hits during simulation period as a function of ws

(continued)

© Springer Nature Singapore Pte Ltd. 2017
N. Mallikharjuna Rao and M. Muniratnam Naidu, *Sliding Window Algorithm for Mobile Communication Networks*, https://doi.org/10.1007/978-981-10-8473-7_4

(continued)

ws	Sliding window size in days
M	$R - g(\text{ws})$, number of misses during simulation period
C_1	Call setup time in case of hit
C_2	Call setup time in case of miss
C	Incremental call setup time due to disk access
AHT	Average hit time
AMT	Average miss time
ACST	Average call setup time
t	Average disk access time
$\bar{r}$	Average number of records in VLR per day

The assumptions of the model are:

1. The call setup time in case of hit, C_1, is constant.
2. The call setup time in case of miss, C_2, is greater than C_1 and constant.
3. The incremental disk access time, C, is varying, and it depends on the number profile records of mobile subscribers in VLR and file structures employed.
4. Disk access time, t, is a constant.

The average hit time and average miss time are given in Eqs. (1) and (2)

$$\text{AHT} = \frac{(C_1 + C)g(\text{ws})}{R} \tag{1}$$

$$\text{AMT} = \frac{C_2(R - g(\text{ws}))}{R} \tag{2}$$

The average call setup time is the sum of average hit time and average miss time given in Eq. (3)

$$\text{ACST} = \frac{(C_1 + C)g(\text{ws}) + C_2(R - g(\text{ws}))}{R} \tag{3}$$

The objective is to minimize average call setup time. Eq. (4) represents the objective function.

$$\text{Minimize } f(\text{ws}) = \frac{(C_1 + C)g(\text{ws})}{R} + \frac{C_2(R - g(\text{ws}))}{R} \tag{4}$$

In the above equation, the parameters, C_1 and C_2, are assigned with 3.0 and 7.5 s, respectively, as per TRAI. The number of call setup requests, R generated during a simulation period for various sliding window sizes, is the same. However, the number of hits and misses during a simulation period depends on the sliding window size.

The incremental disk access time, C, is assumed to be zero up to sliding window of size seven, and it varies depending on file structure employed. The file structures, simple indexed sequential file and B-tree-based multi-level indexed file, are considered [25–27]. It is computed using Eq. (5).

$$C = \begin{cases} 0 \text{ for ws} = 7 \\ t.(\log_2(\bar{r}+1)) \text{ for Simple Indexed Sequential File} \quad \text{for ws} > 7 \\ t.log_m(\bar{r}+1) \text{ for Multi-Level Indexed File of order of m} \quad \text{for ws} > 7 \end{cases}$$

$$(5)$$

The order of B-tree, m, is determined using Eq. (6) such that the depth of B-tree, d, is not more than three levels [26, 29].

$$d \leq 1 + log_{\lceil m/2 \rceil}\left(\frac{\bar{r}}{2}\right) \tag{6}$$

The average disk access, t, depends on the disk characteristics. The TRAI document specifies 0.0216 s as disk access time.

The average number of records in VLR over the simulation period, $\bar{r}$, is computed using Eq. (7)

$$\bar{r} = \frac{\sum_{i=1}^{T} r_i}{T} \tag{7}$$

where

r_i Number of records in VLR at the end of day;
T   Simulation period in days;
$\bar{r}$ Average number of records per day.

The simulation process employed for determining the number of hits and number of disk accesses for a given sliding window size and file structure for evaluating Eq. (4) is presented in the following section.

3   Simulation Process

The input parameter values assumed for performing simulation are given in Table 1.

Table 1 Simulation parameter values

Notations	Description	Values
T	Simulation period in days	1001
ws	Window size	$7 \leq \text{ws} \leq 49$
CP	Calling population size	77,260

The variables used for solving the model through simulation are defined hereunder.

$S_i = \{s | s$ is an MSISDN that makes at least one call setup request on ith day$\}$

$$\text{for}(1 \leq i \leq \text{ws})$$

$$S_i = \{\emptyset\} \text{for } (1 \leq i \leq \text{ws})$$

$$I = \bigcap_{j=1}^{\text{ws}} S_j \quad \text{where ws} = 7, 8, 9, 10, \ldots$$

$$S_{\text{ws}} = S_{\text{ws}} \cup I$$

$$\text{pws} = \text{previous sliding window size}$$

$$\text{ptp} = \text{previous throughput}$$

The algorithm for simulation of OSWS model is as follows:

OSWS $(T, \lambda, \text{CP}, \text{FS})$	
1	ws $\leftarrow$ 7
2	tp $\leftarrow \infty$
3	do
	{
4	ptp $\leftarrow$ tp
5	for $i = 1$ to ws do
6	$S_i \leftarrow \emptyset$
7	VLR $\leftarrow \emptyset$
8	$H \leftarrow 0$
9	CR $\leftarrow 0$
10	$i \leftarrow 1$
11	while$(i \leq \text{ws})$
	{
12	CR $\leftarrow$ Generate-poisson-random-variate(λ)
13	for $j = 1$to CR
	{
14	$s \leftarrow$ Generate-Uniform-Random-Variate(CP)
15	if$(s \in \text{VLR}))$
16	$H \leftarrow H + 1$
	Else
	{
17	VLR $\leftarrow$ VLR $\cup s$
18	$S_i \leftarrow S_i \cup s$
	}

(continued)

(continued)

OSWS $(T, \lambda, \text{CP}, \text{FS})$	
	}
19	$i \leftarrow i + 1$
	}
20	$I = \bigcap\limits_{i=1}^{\text{ws}} S_i$
21	$d \leftarrow 8$
22	while $(d \leq T)$
	{
23	for $i = 1$ to ws $- 1$ do
	{
24	$S_i \leftarrow S_{i+1}$
25	$S_{\text{ws}} \leftarrow I$
	}
26	CR $\leftarrow$ Generate-poisson-random-variate(λ)
27	for $j = 1$ to CR
	{
28	$s \leftarrow$ Generate-Uniform-Random-Variate(CP)
29	if$(s \in \text{VLR}))$
30	$H \leftarrow H + 1$
	else
	{
31	VLR $\leftarrow$ VLR $\cup s$
32	$S_{\text{ws}} \leftarrow S_{\text{ws}} \cup s$
	}
	}
33	$d \leftarrow d + 1$
	}
34	$M \leftarrow \text{CR} - H$
35	$\bar{r} = \text{CR}/T$
36	if FS $= 1$ then
37	$C = t.\log_2(\bar{r} + 1)$
	else
38	$C = t.\log_m(\bar{r} + 1)$
39	acst $\leftarrow (C_1 + C)H + C_2 M/H + M$
40	tp $\leftarrow 1/\text{acst}$
41	pws $\leftarrow$ ws
42	ws $\leftarrow$ ws $+ 1$
	}
43	while(ptp $\geq$ tp)
44	return(pws)

The lines 1–2 initialize the sliding window size and throughput to seven and infinity, respectively. The do-while loop that spans from line 3 to line 43 is executed until the optimal sliding window size is arrived at satisfying the stopping criterion. The lines from 5 to 10 are initialization statements. The while loop that spans from 11 to 19 determines the hits for the first seven days. The function Generate-Poisson-Random-Variate () returns the number of call setup requests for a given day. The function Generate-Uniform-Random-Variate () returns the MSISDN number from calling population source. The statement of line 20 computes the intersection of ws days. The while loop that spans from line 22 to line 33 is executed for a given sliding window size until the termination criterion of simulation is satisfied. The statements from line 34 to line 40 are to compute average call setup time and throughput for a given sliding window size.

4   Simulation Output Analysis

For each file structure, it is simulated for 1001 days and results are given in Tables 2 and 3. Further, the results regard to average hit time (AHT) and average miss time (AMT) and average call setup time (ACST) are shown in Fig. 1 and the results regard to throughput is shown in Figs. 2 through 5.

Table 2 Performance metrics versus sliding window size for simple indexed sequential file

Simulation period (T) = 1001 days							
File structure: *Simple indexed sequential file*							
Window size (ws)	Number of call setup requests (CR)	Average number of records in VLR ($\bar{r}$)	H	Number of disk access	C	ACST	Throughput
7	522	497	423	8.9600	0.1935	3.8347	0.2608
8	603	568	520	9.1523	0.1977	3.5613	0.2808
9	662	639	610	9.3219	0.2014	3.3964	0.2944
10	740	710	690	9.4737	0.2046	3.3256	0.3007
11	819	781	750	9.6110	0.2076	3.3780	0.2960
12	896	852	810	9.7364	0.2103	3.4218	0.2922
13	982	923	870	9.8517	0.2128	3.4590	0.2891
14	1070	994	930	9.9586	0.2151	3.4910	0.2865
15	1142	1065	990	10.0580	0.2173	3.5189	0.2842
16	1213	1136	1050	10.1510	0.2193	3.5433	0.2822
17	1314	1207	1110	10.2384	0.2211	3.5650	0.2805
18	1397	1278	1170	10.3208	0.2229	3.5844	0.2790
19	1475	1349	1230	10.3987	0.2246	3.6018	0.2776

(continued)

Table 2 (continued)

Simulation period (T) = 1001 days

File structure: *Simple indexed sequential file*

Window size (ws)	Number of call setup requests (CR)	Average number of records in VLR ($\bar{r}$)	H	Number of disk access	C	ACST	Throughput
20	1533	1420	1290	10.4727	0.2262	3.6175	0.2764
21	1616	1491	1350	10.5430	0.2277	3.6317	0.2753
22	1694	1562	1410	10.6101	0.2292	3.6448	0.2744
23	1766	1633	1470	10.6742	0.2306	3.6567	0.2735
24	1850	1704	1530	10.7356	0.2319	3.6677	0.2726
25	1940	1775	1590	10.7944	0.2332	3.6779	0.2719
26	2008	1846	1650	10.8510	0.2344	3.6873	0.2712
27	2080	1917	1710	10.9054	0.2356	3.6960	0.2706
28	2142	1988	1770	10.9578	0.2367	3.7042	0.2700
29	2224	2059	1830	11.0084	0.2378	3.7118	0.2694
30	2295	2130	1890	11.0573	0.2388	3.7190	0.2689
31	2358	2201	1950	11.1046	0.2399	3.7257	0.2684
32	2430	2272	2010	11.1504	0.2408	3.7320	0.2680
33	2516	2343	2070	11.1948	0.2418	3.7380	0.2675
34	2600	2414	2130	11.2378	0.2427	3.7436	0.2671
35	2675	2485	2190	11.2796	0.2436	3.7489	0.2667
36	2759	2556	2250	11.3202	0.2445	3.7540	0.2664
37	2822	2627	2310	11.3597	0.2454	3.7588	0.2660
38	2899	2698	2370	11.3982	0.2462	3.7633	0.2657
39	2975	2769	2430	11.4357	0.2470	3.7677	0.2654
40	3042	2840	2490	11.4722	0.2478	3.7718	0.2651
41	3123	2911	2550	11.5078	0.2486	3.7758	0.2648
42	3219	2982	2610	11.5425	0.2493	3.7796	0.2646
43	3290	3053	2670	11.5765	0.2501	3.7832	0.2643
44	3371	3124	2730	11.6096	0.2508	3.7867	0.2641
45	3446	3195	2790	11.6421	0.2515	3.7900	0.2639
46	3522	3266	2850	11.6738	0.2522	3.7932	0.2636
47	3593	3337	2910	11.7048	0.2528	3.7963	0.2634
48	3668	3408	2970	11.7351	0.2535	3.7992	0.2632
49	3753	3479	3030	11.7649	0.2541	3.8021	0.2630

Table 3 Performance metrics versus sliding window size for multi-level indexed file

Simulation period (*T*) = 1001 days

File structure: *Multi-level indexed file*

Window size (ws)	Number of call setup requests (CR)	Average number of records in VLR ($\bar{r}$)	H	Number of disk access	C	ACST	Throughput
7	522	497	423	3.0000	0.0648	3.7252	0.2684
8	603	568	520	3.0000	0.0648	3.4396	0.2907
9	662	639	610	3.0000	0.0648	3.2661	0.3062
10	740	710	690	3.0000	0.0648	3.1897	0.3135
11	819	781	750	3.0000	0.0648	3.2408	0.3086
12	896	852	810	3.0000	0.0648	3.2834	0.3046
13	982	923	870	3.0000	0.0648	3.3195	0.3013
14	1070	994	930	3.0000	0.0648	3.3504	0.2985
15	1142	1065	990	3.0000	0.0648	3.3771	0.2961
16	1213	1136	1050	3.0000	0.0648	3.4006	0.2941
17	1314	1207	1110	3.0000	0.0648	3.4212	0.2923
18	1397	1278	1170	3.0000	0.0648	3.4396	0.2907
19	1475	1349	1230	3.0000	0.0648	3.4560	0.2893
20	1533	1420	1290	3.0000	0.0648	3.4708	0.2881
21	1616	1491	1350	3.0000	0.0648	3.4842	0.2870
22	1694	1562	1410	3.0000	0.0648	3.4964	0.2860
23	1766	1633	1470	3.0000	0.0648	3.5075	0.2851
24	1850	1704	1530	3.0000	0.0648	3.5177	0.2843
25	1940	1775	1590	3.0000	0.0648	3.5271	0.2835
26	2008	1846	1650	3.0000	0.0648	3.5357	0.2828
27	2080	1917	1710	3.0000	0.0648	3.5437	0.2822
28	2142	1988	1770	3.0000	0.0648	3.5512	0.2816
29	2224	2059	1830	3.0000	0.0648	3.5581	0.2811
30	2295	2130	1890	3.0000	0.0648	3.5645	0.2805
31	2358	2201	1950	3.0000	0.0648	3.5706	0.2801
32	2430	2272	2010	3.0000	0.0648	3.5763	0.2796
33	2516	2343	2070	3.0000	0.0648	3.5816	0.2792
34	2600	2414	2130	3.0000	0.0648	3.5866	0.2788
35	2675	2485	2190	3.0000	0.0648	3.5913	0.2784
36	2759	2556	2250	3.0000	0.0648	3.5958	0.2781
37	2822	2627	2310	3.0000	0.0648	3.6000	0.2778
38	2899	2698	2370	3.0000	0.0648	3.6040	0.2775
39	2975	2769	2430	3.0000	0.0648	3.6078	0.2772

(continued)

Table 3 (continued)

Simulation period (T) = 1001 days

File structure: *Multi-level indexed file*

Window size (ws)	Number of call setup requests (CR)	Average number of records in VLR ($\bar{r}$)	H	Number of disk access	C	ACST	Throughput
40	3042	2840	2490	3.0000	0.0648	3.6114	0.2769
41	3123	2911	2550	3.0000	0.0648	3.6148	0.2766
42	3219	2982	2610	3.0000	0.0648	3.6181	0.2764
43	3290	3053	2670	3.0000	0.0648	3.6212	0.2762
44	3371	3124	2730	3.0000	0.0648	3.6242	0.2759
45	3446	3195	2790	3.0000	0.0648	3.6270	0.2757
46	3522	3266	2850	3.0000	0.0648	3.6297	0.2755
47	3593	3337	2910	3.0000	0.0648	3.6323	0.2753
48	3668	3408	2970	3.0000	0.0648	3.6348	0.2751
49	3753	3479	3030	3.0000	0.0648	3.6372	0.2749

Fig. 1 Sliding window size versus AHT, AMT, and ACST

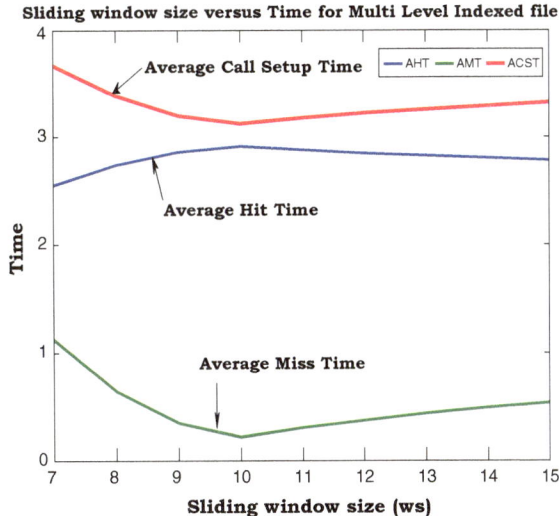

It is evident from Fig. 1 that the behavior of average hit time and average miss time varies as a function of sliding window size is opposing. Obviously, the behavior of the average call setup time as a function of sliding window size is unimodal as shown in Fig. 2. Similarly, the behavior of throughput as a function of sliding window size is unimodal as shown in Fig. 3. The behavior of average call

Fig. 2 Sliding window size versus average call setup time over a range of 7–15

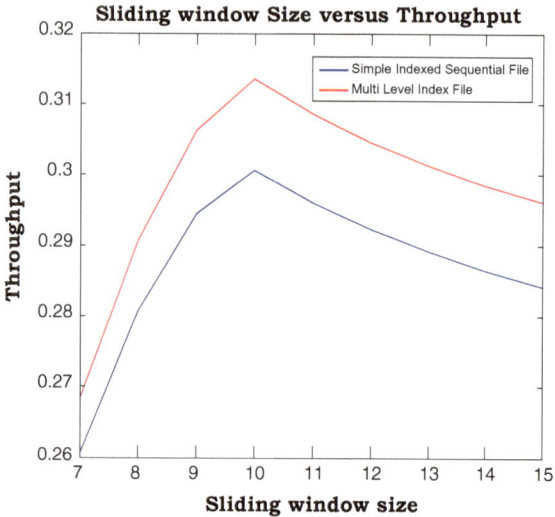

Fig. 3 Sliding window size versus throughput over a range of 7–15

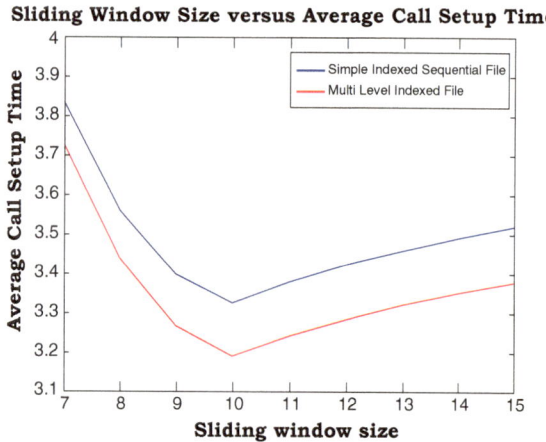

setup time and throughput over a range of 7–49 sliding window sizes is shown in Figs. 4 and 5. The optimal sliding window size that minimizes the average call setup time and maximizes throughput is 10.

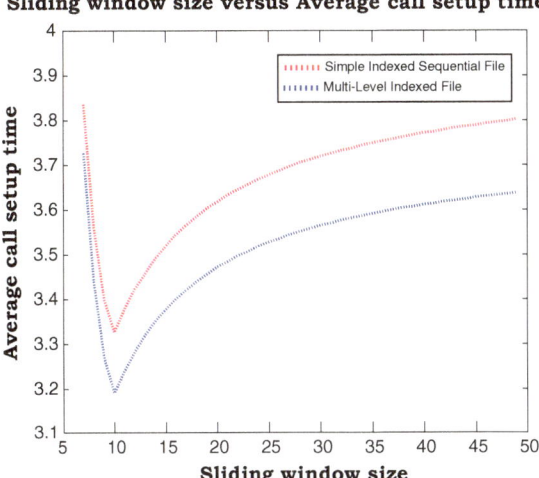

Fig. 4 Sliding window size versus average call setup time over a range of 7–49

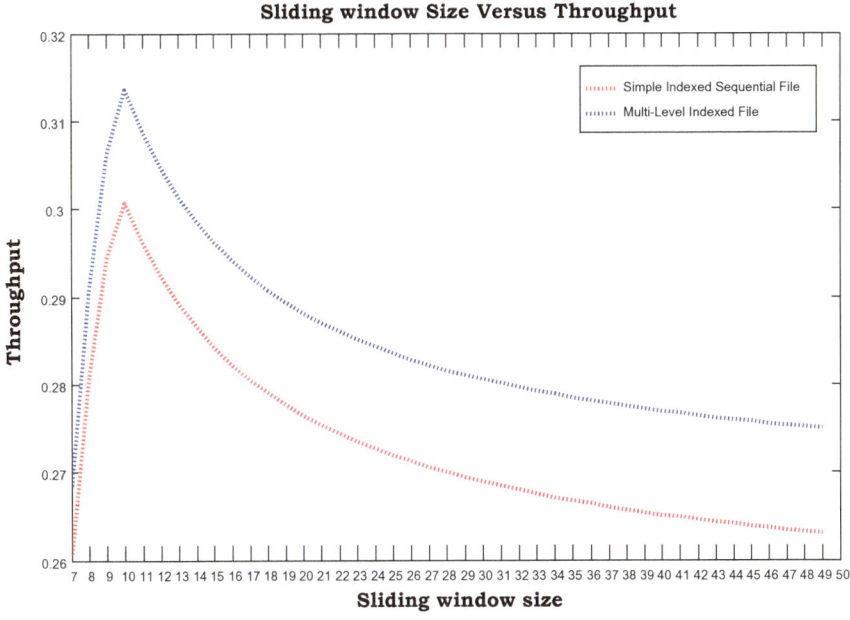

Fig. 5 Sliding window size versus throughput over a range of 7–49

5   Summary

In this chapter, a model is proposed for determining an optimal sliding window size minimizing the average call setup time at an MSC. It is presumed that the objective function behavior is unimodal. However, it cannot be solved either by an analytical method or by a numerical method. Hence, it is solved through simulation. The analysis of simulation output proved that the presumption is valid and the optimal sliding window size is ten for the given input parameters and data assumptions. However, the use of multi-level indexed file structure for VLR represented by B-tree of an appropriate order that does not exceed three levels resulted in lower average call setup time/higher throughput than that of simple indexed file structure. Hence, the model proposed in this chapter is significant for maximizing the throughput of GSM network.

Integrating Sliding Window Algorithm with a Single Server Finite Queuing Model

1 Introduction

In mobile communications, queuing theory is a study about queue as a technique for managing process of call setup request at a Mobile Switching Center. A call setup queue can be studied in terms of: the source of each queued call setup requests, how frequently call setup requests arrive on the queue, how long they can or should wait, whether some call setup requests should jump ahead in the queue, how multiple queues might be formed and managed, and the rules by which call setup request is en-queued and de-queued.

The call setup requests queue that a MSC manages is sometimes viewed as being in stacks. In an MSC, call setup requests are always added to the top of a stack. A process that handles queued call setup requests from the bottom of the stack is known as a first-in, first-out (FIFO) process. A process that handles the call setup requests at the top of the stack first is known as a last-in, first-out (LIFO) process. This chapter discusses how call setup requests are processed in an MSC with a single finite queuing model.

2 A Single Server Finite Queuing Model

An entity that approaches a service facility for want of service is referred to as customer, whereas the service facility is referred to as server. The service on arrival of a customer is commenced immediately provided the server is idle. Otherwise, the customer has to wait in the queue until its turn. The six elements of a queuing model are interarrival, service times of customers, number of servers, queue discipline, maximum queue size, and calling population size.

N. Mallikharjuna Rao and M. Muniratnam Naidu, *Sliding Window Algorithm for Mobile Communication Networks*, https://doi.org/10.1007/978-981-10-8473-7_5

The interarrival and service times of customers are probabilistic in nature following respective probabilistic distributions. The number of servers is either single or multiple. The identical multiple servers are arranged parallelly, whereas the nonidentical multiple servers are arranged in tandem. The queue discipline specifies order in which the next customer is selected from the queue for providing service. The queue disciplines are first-come, first-served (FCFS), last-come, first-served (LCFS), selection in random order (SIRO), and order of priority. The maximum queue size is the maximum number of customers permitted to wait in queue. It is assumed as infinite whenever there is no restriction on queue size. The calling population size is assumed as infinite whenever it exceeds threshold.

A queuing model is descriptive in nature which provides the change of the state of a queuing system due to occurrence of customer arrival and departure events. The state of a queuing system is represented by the status of server (busy or idle), number of customers in system and queue. As the arrival and departure events occur randomly over time, the state of the queuing system changes randomly.

The steady-state performance measures of queuing system server utilization, mean number of customers in system/queue, and mean waiting time in system/queue can be computed. These measures are used in a cost model for deciding the number of servers and quality of service. All setup requests from mobile subscribers are customers, whereas the MSC that can process one request at a time for providing call setup services is a server. Obviously, the number of mobile subscribers is finite. Hence, the queuing system under consideration is formulated as a single server finite queuing model.

3   Integration of Sliding Window Algorithm with a Single Server Finite Queuing Model

The sliding window algorithm ensures minimum average call setup time which is termed as average service time of queuing system. However, the sliding window algorithm does not consider the waiting of call setup request while a MSC is busy.

The average waiting time in queue shall be added to average call setup time to determine the average time a call setup request spends at a MSC. For this purpose, sliding window algorithm is integrated with a single sever finite queuing model. Henceforth, it is referred as an integrated model depicted in Fig. 1. Obviously, the integrated model is a single server finite queuing model. The SWSSD algorithm is employed for processing a call setup requests, which produces two deterministic call setup service times D_h and D_m as a consequence of hit and miss respectively. Therefore, the Kendall-Lee notation for representing the model is as follows:

$$M/D/1 : (\text{FCFS}/\infty/\infty)$$

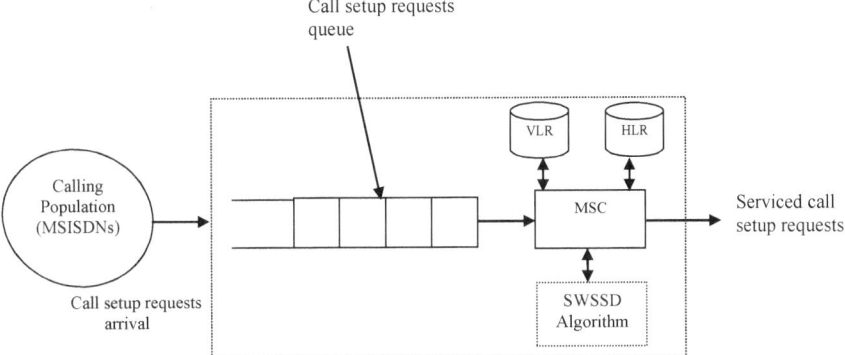

Fig. 1 Integrated model

It is assumed that the customer (call setup request) arrival rate follows Poisson distribution with parameter λ denoted by M whereas D is a deterministic service time (call setup time) as defined below:

$$D = \begin{cases} D_h, & \text{if record is available in VLR (hit)} \\ D_m, & \text{otherwise (miss)} \end{cases}$$

Obviously, an analytical solution cannot be found for computing average waiting time of call setup requests in queue and it is to be determined through simulation.

4 Simulation

The simulation of the integrated model generates an artificial history of system facilitating the measurement of throughput of a MSC. The entities considered for developing simulation model are mobile stations and Mobile Switching Center. The mobile stations are represented by their unique identities, MSISDNs. The call setup requests from MSISDNs are the customers. The Mobile Switching Center (MSC) is a server. A set of mobile stations registered with a service provider constitute the calling population source from which call setup requests are received.

In GSM network, the profiles of MSISDNs comprising several attributes are stored in HLR at Gateway Mobile Switching Center (GMSC). The profiles of MSISDNs in the service area of a MSC are stored in VLR. Whenever an MSISDN profile is not available in VLR for processing its call setup request, its profile is fetched from HLR. The deletion of the profile of an MSISDN from VLR depends on the sliding window algorithm.

The metric such as throughput is employed for evaluating the performance of a MSC using integrated model is throughput considering waiting time in queue would be more realistic. Equations (1) and (2) define the average waiting time in the system and throughput, respectively.

$$AWTS = \frac{W_q + D}{\text{No. of call setup requests}} \tag{1}$$

where

$$AWTS = \text{Average waiting time in system}$$
$$W_q = \text{average waiting time in queue}$$
$$D = \text{is a deterministric servie time}$$
$$D = \begin{cases} D_h, & \text{if record is available in VLR} \\ D_m, & \text{otherwise} \end{cases}$$

Obviously, D_m is greater than D_h as the relevant record is to be fetched from HLR in case of its non-availability. From the analysis of data pertaining to call setup times of Telecom Regulatory Authority of India (TRAI), D_h and D_m are assigned 3 and 7 s, respectively.

The reciprocal of average waiting time in system is throughput as shown in Eq. (2)

$$\text{Throughput (TP)} = \frac{1}{\text{Average waiting time in system}} \tag{2}$$

The simulation is performed using TN3270 emulator for IBM (zSeries) mainframe on Windows platform with z/OS IBM mainframe operating system and DB2 database server. The notations and input parameters of simulation are shown in Table 1.

The output of simulation for 88,243 call setup requests over 1001 days is presented in Table 2. A call setup request has three attributes, viz., calls per day,

Table 1 Simulation parameters

Symbol	Description	Value
N	Simulation run length in days	1001
M	Calling population size	88,243
C	Ordered set of calling population	$\{MSISDN_i \in \mid 1 \leq i \leq 88{,}243\}$
λ	Parameter of Poisson distribution, i.e., mean number of call setup requests per day	87
$1/\mu$	Parameter of exponential distribution	0.02
D_h	Call setup time in case of record available in VLR	3
D_m	Call setup time in case of record unavailable in VLR	7
WS	Sliding window size in days	7

Table 2 Simulation results

Call setup request no (i)	Call setup request attribute values			Service time attribute values			Arrival time (AT_i)	Service begins (SB_i)	Service ends (SE_i)	Waiting time in queue (WTQ_i)	Waiting time in system (WTS_i)
	Day (D_i)	MSISDN	Interarrival time (IAT_i)	Hit (Yes/ No) (SIDA_i)	Call setup time (ST_i)						
(1)	(2)	(3)	(4)	(5)	(6)		(7)	(8)	(9)	(10)	(11)
1	1	65	93	No	7		0	0	7	0	7
2	1	73	41	No	7		93	93	100	7	14
3	1	23	22	No	7		134	134	141	100	107
4	1	47	48	No	7		156	156	163	141	148
5	1	74	87	No	7		204	204	211	163	171
⋮	⋮	⋮	⋮	⋮	⋮		⋮	⋮	⋮	⋮	⋮
⋮	⋮	⋮	⋮	⋮	⋮		⋮	⋮	⋮	⋮	⋮
88,242	1001	41	5	Yes	3		3298	3298	3301	3284	3287
88,243	1001	17	29	No	7		3303	3303	3310	3301	3308

MSISDN requesting for call setup, and interarrival (time between two successive call setup requests) as shown in columns 2, 3, and 4, respectively. The number of call setup requests per day is a random variate from Poisson distribution with parameter $\lambda = 87$.

A discrete uniform random variate represents the attribute MSISDN, and its interarrival time is a random variate from an exponential distribution with parameter $1/\mu = 0.02$.

The SWSSD algorithm returns 'Yes' if it is a hit, otherwise 'No' as shown in column 5. Accordingly, call setup service times are 3 and 7 for hit and miss, respectively, as shown in column 6.

$$AT_i = \begin{cases} 0 & \text{for } i = 1 \\ AT_{i-1} + IAT_i & \text{for } i > 1 \end{cases} \tag{3}$$

$$SB_i = \text{Max}(AT_i, SB_{i-1}) \tag{4}$$

$$SE_i = SB_i + ST_i \tag{5}$$

$$WTQ_i = SB_i - AT_i \tag{6}$$

$$WTS_i = SE_i - AT_i \tag{7}$$

The arrival time (AT_i) with respect to each call setup request is computed using Eq. (3) as shown in column 7. Service Beginning Time (SB_i) and Service Ending Time (SE_i) with respect to each call setup request are computed using Eqs. (4) and (5), respectively, as shown in columns 8 and 9.

The waiting time in queue (WTQ_i) and waiting time in system (WTS_i) for each of 88,243 call setup requests over 1001 days are computed and represented graphically in Figs. 2, 3, and 4, respectively.

$$AWTQ = \frac{\sum WTQ_i}{N} \tag{8}$$

$$AWTS = \frac{\sum WTS_i}{N} \tag{9}$$

Further, the numbers of hits are determined for each of 143 blocks when the call setup requests are processed by employing sliding window of size seven days algorithm [23] for over a period of 1001 days. This model partitions the sliding window into seven blocks and maps onto seven days. The call setup request for each block of seven days is aggregated. The performance metrics such as hits, average call setup time, and throughput are computed for each block, and the same is briefed in Table 3. When the call setup requests are processed by employing integrated model, the waiting time in queue and service times is determined for each of 143 blocks. Performance metrics such as average waiting time in queue, average

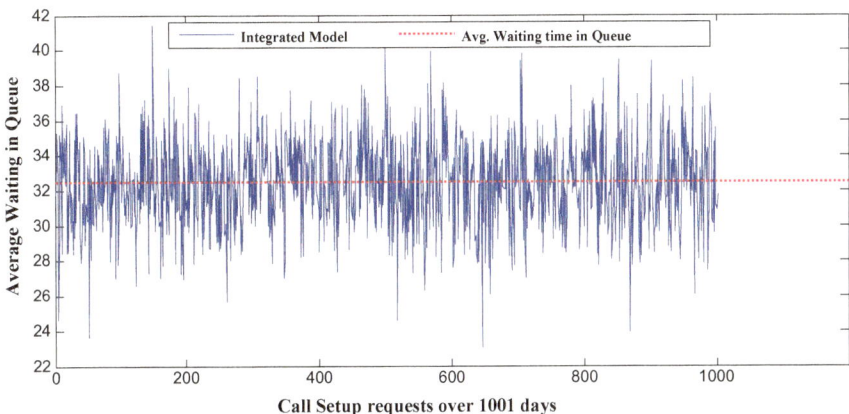

Fig. 2   Waiting time in system over 1001 days

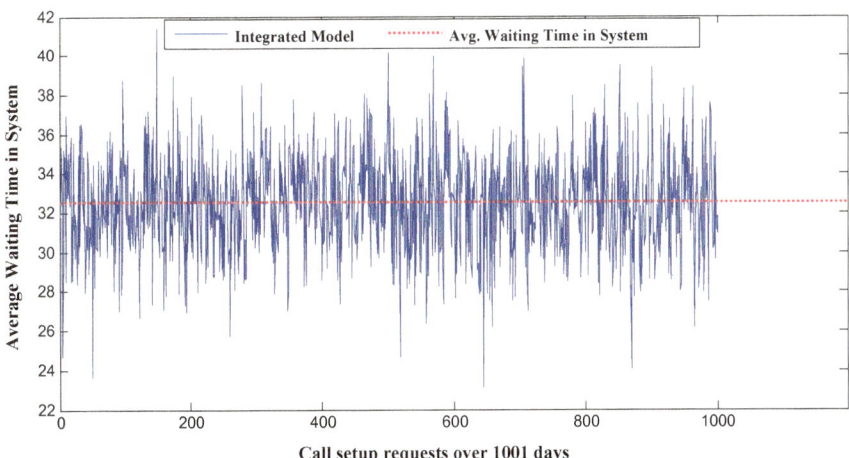

Fig. 3   Waiting time in queue over 1001 days

waiting time in system, and realistic throughput are presented in Table 3, and the corresponding graphical representations are shown in Figs. 5 and 6.

The performance metrics such as average waiting time in queue (AWTQ) and average waiting time in system (AWTS) computed using Eqs. (8) and (9) are 32.4552 and 32.5062, respectively. Accordingly, the throughput of a MSC is 0.03097 which is a reciprocal of AWTS, whereas the throughput of SWSSD algorithm without considering waiting in queue at a MSC is 0.18345.

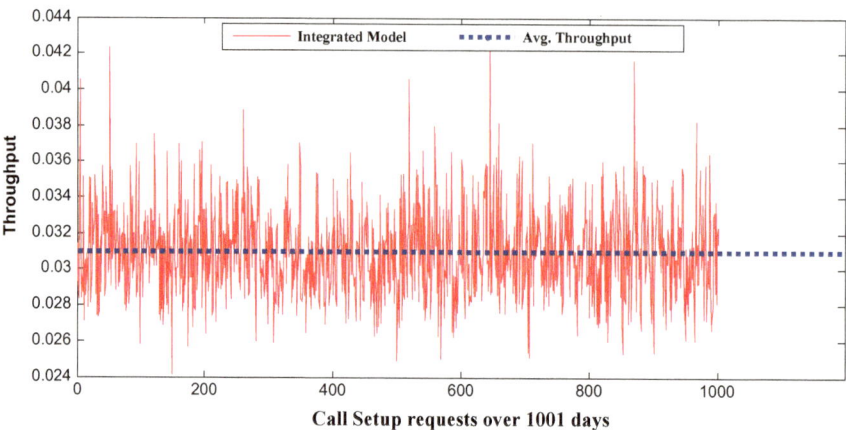

Fig. 4 Throughput over 1001 days

Table 3 Aggregated simulation results for each block

Period/ blocks	Call setup requests	Hits	Sliding window algorithm		Integrated model		
			Average call setup time	Throughput	Average waiting time in queue (AWTQ)	Average waiting time in system (AWTS)	Throughput
1–7	621	318	4.951691	0.2019	31.8776	31.9178	0.03133
8–14	549	262	5.091075	0.1964	39.1165	39.1621	0.02553
15–21	631	239	5.484945	0.1823	32.0570	32.1030	0.03115
22–28	616	244	5.415584	0.1846	31.5487	31.5957	0.03165
…	…	…	…	…	…	…	…
…	…	…	…	…	…	…	…
988–994	638	241	5.489028	0.1821	33.5109	33.5501	0.02806
995–1001	629	206	5.689984	0.1757	31.9793	32.0445	0.03120

It is evident that the throughput of the proposed integrated model which considers waiting time in queue at a MSC is decreased by 83.12% for a single channel MSC; it is a realistic measurement of throughput.

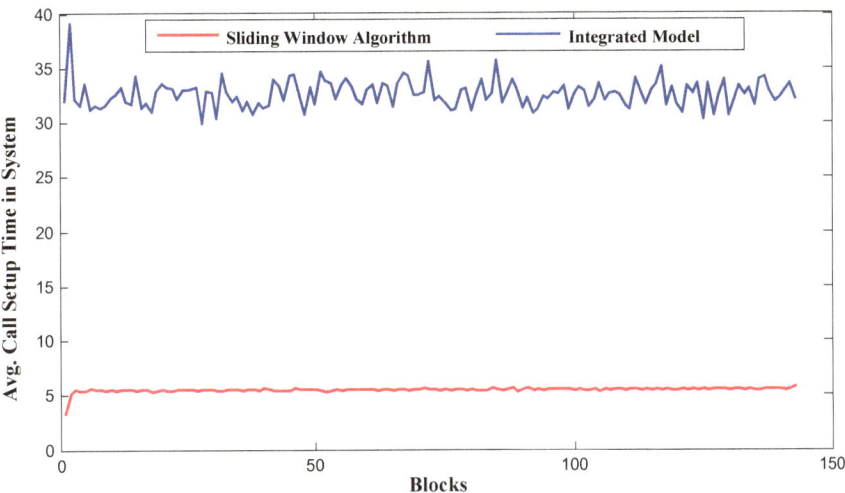

Fig. 5 Call setup time in system versus blocks at a MSC

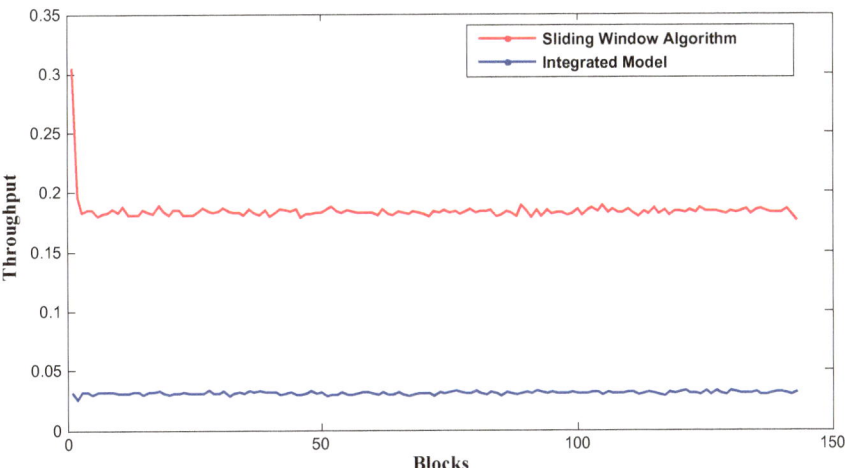

Fig. 6 Throughput versus blocks at a MSC

5   Summary

In this chapter presented a realistic model for measuring throughput of a MSC integrating the sliding window algorithm with a single server finite queuing model. The sliding window algorithm minimizes the average call setup processing time and a single server finite queuing model determines the average call setup request waiting time in the queue. Hence, the reciprocal of the sum of average call setup processing time and average call setup request waiting time in the queue is more realistic average throughput of a MSC under consideration. This model facilitates to determine an optimal sliding window size.

However, the model assumes a MSC with a single channel for processing call setup requests. In the next chapter, developed a model considering a MSC with multiple identical channels for processing concurrently multiple call setup requests for measuring still more realistic throughput of a MSC.

Integrating Sliding Window Algorithm with a Multiple Server Finite Queuing Model

1 Introduction

The next logical step is to look at a multichannel queuing system, in which two or more servers or channels are available to handle arriving mobile subscribers call setup requests. Let us assume that call setup requests are awaiting service form one single line and then proceed to the first available server at a MSC.

In the previous chapter of this book, a simulation model integrating sliding window algorithm with a single server finite queuing model for measuring the throughput of a MSC is presented. The data generated through simulation is analyzed and found that the throughput of a MSC is reduced by 83.12%. This chapter discuss a simulation model integrating sliding window algorithm with a multi-channel finite queuing model assuming that more than one call setup request can be processed concurrently by a MSC.

2 Multiple Channel Finite Queuing Model

An entity that approaches a service facility for a want of service is referred to as call setup request whereas the service facility is referred to as channel/server. The service on arrival of call setup request is commenced immediately provided the channel is idle. Otherwise, the call setup request has to wait in the queue till its turn. It has six elements, viz., interarrival time, service times of call setup requests, number of channels, queue discipline, queue size, and finite calling population size. The behavior and relations among these six elements of the queue are described in detail in [30].

© Springer Nature Singapore Pte Ltd. 2017
N. Mallikharjuna Rao and M. Muniratnam Naidu, *Sliding Window Algorithm for Mobile Communication Networks*, https://doi.org/10.1007/978-981-10-8473-7_6

The steady-state performance measures of queuing system channel utilization, mean number of call setup requests in system/queue, and mean waiting time in system/queue can be computed. These measures are used in a cost model for determining the number of channels and quality of service. All call setup requests from mobile subscribers are service requests, whereas MSC can process multiple call setup requests concurrently at time. Therefore, the number of call setup requests is finite. Hence, the queuing system under consideration is formulated as a multiple channel finite queuing model. This model includes a single waiting line and service facility with two or more independent channels concurrently.

3 Integrating Sliding Window Algorithm with a Multiple Channel Finite Queuing Model

The sliding window algorithm ensures minimum average call setup time which is termed as average service time of queuing system. However, the sliding window algorithm does not consider the waiting time of call setup request while MSC is busy. The average waiting time in queue shall be added to average call setup time to determine the average time a call setup request spends at MSC. For this reason, sliding window algorithm is integrated with a multiple channel finite queuing model. Henceforth, it is referred as an integrated model with a multiple channel as shown in Fig. 1. Therefore, the integrated model with a multiple channel is a multiple channel finite queuing model.

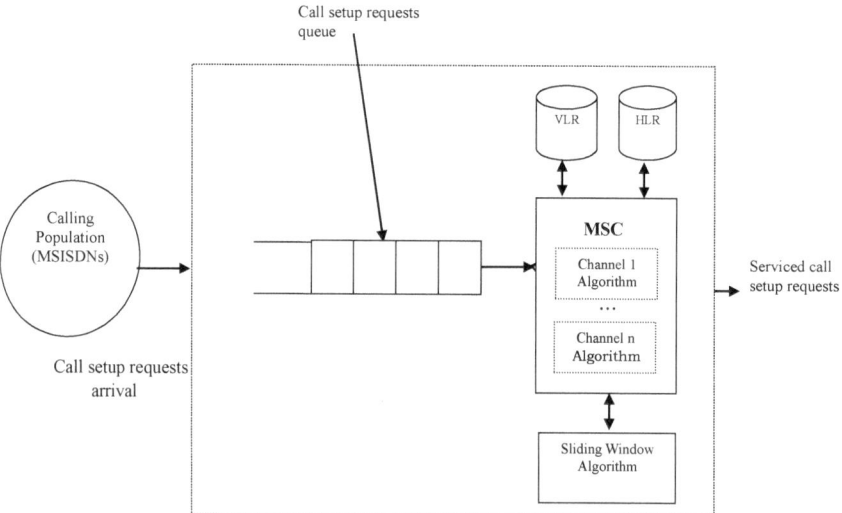

Fig. 1 Integrated model with a multiple channel

The sliding window algorithm is employed for processing a call setup requests, which produces two deterministic call setup service times S_h and S_m as a consequence of hit and miss respectively. Therefore, the Kendall-Lee notation for representing the proposed model is as follows:

$$R/S/k : \text{FCFS}/\infty/\infty$$

It is assumed that the call setup requests arrival rate follows Poisson distribution with parameter λ (mean arrival rate) denoted by R whereas S is a deterministic service time as defined below.

$$S = \begin{cases} S_h, \text{call setup time in case of hit} \\ S_m, \text{call setup time in case of miss} \end{cases}$$

From the analysis of data pertaining to call setup times of Telecom Regulatory Authority of India (TRAI), S_h and S_m are assigned 3 and 7 s, respectively. Obviously, $S_m > S_h$ and also assumed that the number of channels is k, the queuing discipline is first-come, first-served (FCFS) and with finite length of call setup requests. Hence, an analytical solution cannot be found for computing average waiting time of call setup requests considering that call setup requests are in queue and it is to be achieved through computer simulation.

4 Simulation Output Analysis

The simulation of the integrated model with a multiple channel model generates synthetic data facilitating the measurement of throughput of a multiple channel MSC which can process concurrently call setup requests. Two entities considered for developing simulation model are mobile stations and MSC. The mobile stations are represented by their unique identities, viz., MSISDNs. The call setup requests (Calling population) from MSISDNs treating them as a customers in queuing system. A set of mobile stations register with a service provider constitute the calling population source from which call setup requests are received.

In GSM network, the records of MSISDNs consisting set of attributes are stored in HLR. The records of MSISDNs in the service area of a MSC are stored in VLR. Whenever MSISDN initiates a call setup request, its relevant record must be available in VLR of a MSC for routing its call setup request. If it is not available in VLR, its record is fetched from HLR. The deletion of the relevant record of an MSISDN from VLR depends on sliding window algorithm.

The metrics such as average call setup time and throughput are employed for evaluating the performance of a MSC using integrated model with Multiple Channel is throughput considering waiting time in queue would be more realistic. Equations (1) and (2) define the average waiting time in the system and throughput respectively.

$$\text{AWTS} = \frac{W_{\text{queue}} + S}{\text{CR}} \tag{1}$$

where

AWTS   Average wating time in system
W_{queue} Average waiting time in queue
S Deterministic service time
CR   No. of call setup requests

The reciprocal of average waiting time in system is throughput as shown in Eq. (2)

$$\text{Throughput} = \frac{1}{\text{AWTS}} \tag{2}$$

The simulation is performed using TN3270 emulator for IBM mainframes on Windows platform with z/OS IBM mainframe operating system and DB2 database server. The notations and input parameters of simulation are shown in Table 1.

The outcome of simulation for 76,025 call setup requests over 1001 days is presented in Table 2. A call setup requests has three attributes such as calls per day, MSISDN requesting for call setup, and interarrival time as shown in columns 2, 3, and 4, respectively. The number of call setup requests per day is random variate from Poisson distribution with parameter $\lambda = 78$. A discrete uniform random variate represents the attribute MSISDN, and its interarrival time is random variate from an exponential distribution with parameter $\frac{1}{\mu} = 0.02$. The sliding window algorithm returns 'Yes' if it is a hit, otherwise 'No' if it is a miss, as shown in column 5. Hence, the call setup times are 3 and 7 for hit and miss, respectively, as shown in column 6.

$$\text{AT}_i = \begin{cases} 0 & \text{for } i = 1 \\ \text{AT}_{i-1} + \text{IAT}_i & \text{for } i > 1 \end{cases} \tag{3}$$

Table 1 Input/output simulation parameters

Symbol	Description	Value
T	Simulation run length in days	1001
M	Calling population size	76,025
C	Ordered set of calling population	$\{\text{MSISDN}_i \in \lvert 1 \leq i \leq 76025\}$
λ	Parameter of Poisson distribution, i.e., mean number of call setup requests per day	78
$1/\mu$	Parameter of exponential distribution	0.02
S_h	Call setup time in case of record available in VLR	3
S_m	Call setup time in case of record unavailable in VLR	7

Table 2 Simulation results

| Call setup request no (i) | Call setup request Attribute values | | Interarrival time (IAT$_i$) | Service time Attribute values | | Arrival time (AT$_i$) | Channel 1 | | Channel 2 | | Waiting time in queue (WTQ$_i$) | Waiting time in system (WTS$_i$) |
	Day (D_i)	MSISDN		Hit (Yes/No)	Call setup time			Service ends (C$_1$SE$_i$)	Service begins (C$_2$SB$_i$)	Service ends (C$_2$SE$_i$)		
(1)	(2)	(3)	(4)	(5)	(6)	(7)	(8)	(9)	(10)	(11)	(12)	(13)
1	1	70	12	No	7	0	0	7			0	7
2	1	43	7	No	7	12	12	19			7	14
3	1	16	1	No	7	19	19	26			19	26
4	1	15	1	No	7	20			20	27	20	27
5	1	76	15	No	7	21	26	33			26	33
...	...	...	...	...	...	...	...	...	...	...	...	...
...	...	...	...	...	...	...	...	...	...	...	...	...
76,024	1001	10	2	Yes	3	805	805	808			805	808
76,025	1001	14	14	No	3	807			807	810	807	810

$$\text{if } (C_1\text{SE}_{i-1} > \text{AT}_i)$$
$$C_2\text{SB}_i = \text{Min}(\text{AT}_i, C_2\text{SE}_i)$$
$$C_2\text{SE}_i = C_2\text{SB}_i + ST_i \tag{4}$$
$$\text{else } C_1\text{SB}_i = \text{Max}(\text{AT}_i, C_1\text{SB}_{i-1})$$
$$C_1\text{SE}_i = C_1\text{SB}_i + ST_i$$

$$\text{WTQ}_i = \text{SB}_i - \text{AT}_i \tag{5}$$

$$\text{WTS}_i = \text{SE}_i - \text{AT}_i \tag{6}$$

The arrival time (AT_i) with respect to each call setup request is computed using Eq. (3) as shown in column 7.

Service Beginning Time, $(C_1\text{SB}_i)$ and $(C_2\text{SB}_i)$, and Service Ending Time, $(C_1\text{SE}_i)$ and $(C_2\text{SE}_i)$, with respect to each call setup requests in channel 1 and channel 2 are computed using Eq. (4), and values are shown in column 8, 9, 10, and 11, respectively. Call setup requests are performed by multiple channels basing on idle state of the channel. Call setup requests are shared among the channels simultaneously. The waiting time in queue (WTQ_i) and waiting time in system (WTS_i) for each of 76,025 call setup requests over 1001 days are computed using Eqs. (5) and (6) and are presented in columns 12 and 13, respectively.

$$\text{AWTQ} = \frac{\sum \text{WTQ}_i}{T} \tag{7}$$

$$\text{AWTS} = \frac{\sum \text{WTS}_i}{T} \tag{8}$$

The performance metrics such as average waiting time in queue (AWTQ) and average waiting time in system (AWTS) are computed using Eqs. (7) and (8) considering a multiple channel are 9.7515 and 9.9698 respectively. Accordingly, the throughput of a MSC is 0.1009 which is reciprocal of AWTS, whereas the throughput of sliding window algorithm considering single server finite model is 0.09627, and the throughput of sliding window algorithm without considering waiting in queue at a MSC is 0.1842.

Further, the number of hits is computed for each of 143 blocks when the call setup requests are processed by employing sliding window algorithm for over a period of 1001 days. Here, the model has partitioned the sliding window into block and maps onto seven days. The call setup requests for each block of seven days are aggregated.

The performance metrics such as hits, average call setup time, and throughput are computed for each block, and the same is briefed in Table 3. When the call setup requests are processed by employing integrated model and integrated model with a multiple channel, the waiting time in queue and service times is determined for each of 143 blocks. Performance metrics such as average waiting time in queue,

Table 3 Aggregated simulation results for each block

Period/ blocks	Call setup requests	Hits	Sliding window algorithm		Integrated model			Integrated model with a multiple channel		
			Average call setup time	Throughput	Average waiting time in queue (AWTQ)	Average waiting time in system (AWTS)	Throughput	Average waiting time in queue (AWTQ)	Average waiting time in system (AWTS)	Throughput
1–7	536	447	3.6642	0.2729	9.9982	10.1661	0.0983	9.0256	9.3321	0.1072
8–14	545	200	5.5321	0.1808	10.1232	10.4220	0.0960	9.1256	9.6753	0.1034
15–21	518	214	5.3475	0.1870	9.9852	10.2317	0.0978	9.2145	9.6988	0.1031
22–28	552	220	5.4058	0.1850	10.2542	10.6740	0.0937	9.9985	10.1974	0.0981
...	...	...	...	...	...	...	...	...	...	...
...	...	...	...	...	...	...	...	...	...	...
988–994	518	185	5.5714	0.1794	10.8956	11.0599	0.09042	10.0123	10.4498	0.0957
995–1001	550	216	5.4291	0.1842	10.0025	10.3764	0.0964	9.8963	10.0855	0.0992

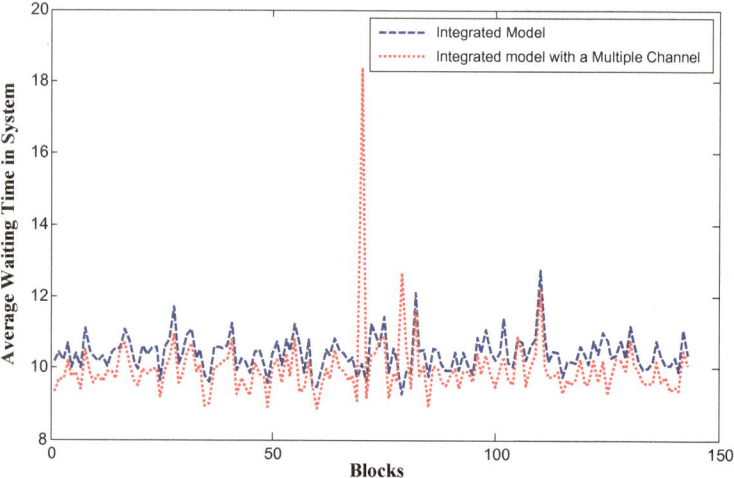

Fig. 2 Average waiting time in system versus blocks at a MSC

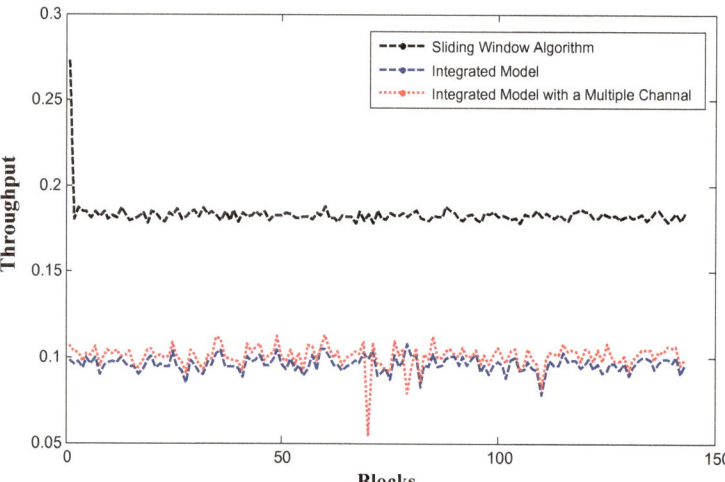

Fig. 3 Throughput versus blocks at a MSC

average waiting time in system, and realistic throughput are presented in Table 3. Graphical representations of average waiting time in system and throughput are shown in Figs. 2 and 3. It is obviously evident from Table 4 that the performance of the proposed integrated model with a multiple channel is better than the integrated model [30].

It is used to generate the performance measures of integrated model and integrated model with multiple channel over a period of 1001 days. It is evident from

Table 4 Comparison of performance metrics

Performance metric	Integrated model	Integrated model with a multiple channel	Increase (%)
Average waiting time in queue	10.0322	9.5250	5.33
Average waiting time in system	10.4097	9.9698	1.41
Average throughput	0.0963	0.1009	4.78

Table 4 that there is significant increase in the performance metrics of the proposed integrated model with a multiple channel for one MSC service area. Obviously, it is recommended to consider adopting the integrated model with a multiple channel model for the entire GSM network for improving its throughput by 4.78%

5 Summary

The proposed model, integrating the sliding window algorithm with a multiple channel finite queuing model for measuring the throughput of a MSC which can process call setup requests concurrently, is realistic. The experimental results of sliding window algorithm integrating with two-channel finite queuing model proved it, as the average throughput is increased by 4.78%. It is obvious that increasing channels of a MSC would improve its throughput for a given sliding window size. The insights provided by the model [30] for determining an optimal sliding window size and the proposed model are immensely useful for formulating a cost model to determine an optimal window size as well as an optimal number of channels for maximizing the throughput of a MSC. Consequently, the present study provides scope for further research.

Method for Determining Optimal Number of Channels

1 Introduction

A cost model is discussed for determining an optimal number of channels minimizing the average call setup time and maximize throughput at a MSC. It is assumed that the objective of proposed cost model behavior is unimodal. However, it cannot be solved either by an analytical or a numerical method. Therefore, it is solved through simulation.

The model discussed in this chapter is for measuring throughput of a Mobile Switching Center (MSC) integrating the sliding window algorithm with a single server finite queuing model referred to as integrated model (IM) [30], which is nearer to the real-time situation as it considered the waiting times of call setup requests. However, in the real-time situation, a MSC can process call setup requests concurrently. In [31] proposed more realistic model to integrate the sliding window algorithm with a multiple channel finite queuing model for measuring the throughput of a MSC. Both models are validated using a simulation model developed for that purpose.

As further study, it is attempted to investigate the behavior of average waiting time in system as a function of sliding window size with multiple channels through simulation model. As sliding window size increases, the number of hits increases, because the chance of availability of a record in VLR also increases. Therefore, record access time and the waiting time of call setup requests also increases. It is observed that when the window size increases cost for average waiting time in the system increases and subsequently decreases. At the same time, the cost of average waiting time decreases with the increase in channel size. Hence, it is concluded that there exists an optimal number channels.

$$\text{maximize} \quad tp = f(sws, nc)$$

© Springer Nature Singapore Pte Ltd. 2017
N. Mallikharjuna Rao and M. Muniratnam Naidu, *Sliding Window Algorithm for Mobile Communication Networks*, https://doi.org/10.1007/978-981-10-8473-7_7

where

sws   is sliding window size and
nc   is number of channels.

Sliding window size versus number of channels

Window size	Channels				
	$Channel_1$	$Channel_2$	$Channel_3$	...	$Channel_n$
7					
8					
9					
10					

2   The Model

The notation and assumptions of the cost model are given below.

Notation:

tp	Throughput
sws	Sliding window size
nc	Number of channels
T	Simulation period in days
R	Number of call setup requests
WTS	Waiting time in system
TWTS	Total waiting time in system
WTQ	Waiting time in queue
D	Cost of call setup time is a deterministic service time
C_1	Cost of call setup time in case of record availability in VLR
C_2	Cost of call setup time in case of record non-availability
C	Incremental cost of call setup time due to disk access
C_c	Cost per channel
C_w	Cost of waiting time in queue
AWTS	Average waiting time in system

The assumptions of the model are:

1. The call setup time in case of record availability, C_1 is constant.
2. The call setup time in case of record non-availability, $C_2 > C_1$ and it is constant.
3. The incremental call setup cost, C is varying and it depends on the number profiles of mobile subscriber in VLR and file structure employed.
4. Disk access time, t is a constant.

The waiting time in system is given in Eq. (1)

$$\text{Waiting Time in System (WTS)} = \text{WTQ} + D \tag{1}$$

where WTQ = waiting time in queue and D is a deterministic call setup time which is obtained from Eq. (2)

$$D = \begin{cases} C_1 & \text{if record is available in VLR} \\ C_2, & \text{Otherwise} \end{cases} \tag{2}$$

Therefore, waiting time in system is obtained by using Eq. (3)

$$\text{TWTS} = (C + \text{WTS}) \tag{3}$$

The incremental cost of call setup time, C is assumed to be zero up to sliding window of size seven and it varies on file structure employed. Simple indexed file and B-tree-based multi-level indexed file are considered. It is computed using Eq. (4)

$$C = \begin{cases} 0 & \text{for } ws = 7 \\ t.(\log_2(\bar{r} + 1)) \text{ for Simple Indexed Sequential File} & \text{for } ws > 7 \\ t.\log_m(\bar{r} + 1) \quad \text{for Multi-Level Indexed File of order of } m & \text{for } ws > 7 \end{cases}$$
$$\tag{4}$$

The average disk access t depends on the disk characteristics. The Telecom Regulatory Authority of India (TRAI) document specifies 0.0216 s as disk access time.

The simulation process employed for determining the average waiting time in system for a given sliding window size and considering multiple channels.

In order to determine cost of optimal number of channels with the total optimal cost of the system, two opposing costs must be considered. The solution to a multichannel queuing problem may require management to make a trade-off between the increased cost of providing better service and the decreased waiting costs derived from providing that service. These are cost of service and cost of waiting time in queue:

$$\text{Cost of service} = (\text{Number of channels}) * (\text{Cost per channel})$$

$$\text{Cost of service } (C_s) = nc * C_c$$

The waiting cost when the waiting time cost which is based on time in the system is

$$\text{Cost of waiting time in queue} = (\text{Total time spent waiting by all call setup requests})$$
$$(\text{Cost of waiting})$$

$$\text{Cost of waiting time in queue } \left(C_{wq}\right) = (\text{TWTS})(C_w)$$

Adding the cost of service to the total cost of waiting time in queue, we have the total cost of the queuing system. When the waiting cost based on the time in the system, this is

$$\text{Total Cost} = \text{cost of Service} + \text{cost of waiting time in queue}$$

$$\text{Total Cost} = \left(C_s\right) + \left(C_{wq}\right)$$

The feasibility of the cost model depends on the estimation of cost parameters. These parameters are difficult to estimate mostly the one associated with the waiting time of call setup requests in a queue at an MSC. To avoid this difficulty, we used an aspiration-level model which attempts to minimize this difficulty by working directly with the measures of performance of the queue situation.

This study demonstrates the procedure by applying it to the multiple channel model, where it is desired to determine an optimal number of channels, by considering the following measures of performance

1. The average waiting time in system, AWTS
2. The idleness percentage of the channels, $C*$

The idleness percentage of the channels can be computed as follows:

$$C^* = \frac{C - \overline{C}}{C} \times 100$$

where

C *Number of channels*
$\overline{C}$ Number of idleness channels

$$C^* = \frac{C - \left(L_s - L_q\right)}{C} \times 100$$

where

L_s no of call setup request or length of call setup requests is system
L_q length of call setup request in queue.

3   Simulation Process

The simulation is performed using TN3270 emulator for IBM (ZSeries) mainframe on windows platform with z/OS IBM mainframe operating system and DB2 database server. The notations and input parameters of simulation are shown in Table 1.

For each channel, it is simulated for 1001 days and results are shown in Table 2. The output of simulation for 98243 call setup requests over 1001 days is shown in Table 2. A call setup request has three attributes, viz., calls per day, MSISDN requesting for call setup, and interarrival (time between two successive call setup requests).

It is evident from the results, it is concluded that as the service level increase at an optimal service level, the waiting time of the call setup time is reduced. The service efficiency is increased and the mobile subscriber satisfaction is increased and as well as it is found that at some specific number of channels that system cost is minimized and throughput of the system is maximized as shown in Fig. 1.

The analysis of simulation output proved that the presumption is valid, and the optimal number of channel is four for the given input parameters and data assumptions. However, the use aspiration-level model at VLR, resulted lower average calls setup time/higher throughput. Hence, the model discussed in this chapter is significant for maximizing the throughput of GSM network.

Table 1 Simulation parameters

Symbol	Description	Value
N	Simulation run length in days	1001
M	Calling population size	98,243
C	Ordered set of calling population	$\{\text{MSISDN}_i \in \|1 \leq i \leq 98243\}$
λ	Parameter of Poisson distribution, i.e., mean number of call setup requests per day	87
$1/\mu$	Parameter of exponential distribution	0.02
D_h	Call setup time in case of record available in VLR	3
D_m	Call setup time in case of record unavailable in VLR	7
WS	Sliding window size in days	7

Table 2 Average waiting time with channel idleness

	No. of channels											
	1	2	3	4	5	6	7	8	9	10	11	12
Average waiting time in system (AWTS)	99	85	81	76	69	59	51	43	39	31	29	21
Channel idleness Percentage (C*)	0	50	66.66	75	80	83.33	85.71	87.5	88.89	90	90.90	91.22

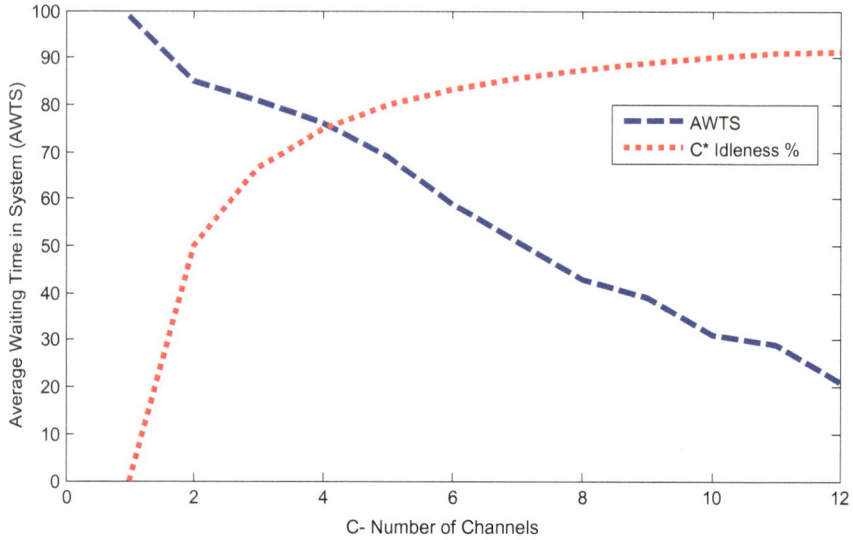

Fig. 1 Number of channels versus average waiting time in system

4   Summary

In this chapter, a simulation process is discussed for determining the average waiting time in system for a given sliding window size and considering multiple channels. An aspiration-level model used which attempts to minimize this difficulty by working directly with the measures of performance of the queue situation.

The analysis of simulation output proved that the presumption is valid, and the optimal number of channel is four for the given input parameters and data assumptions. Hence, the model discussed in this chapter is significant for maximizing the throughput of GSM network.

The aspiration model proposed in this chapter is useful for determining the optimal number of channels for a given sliding window size to satisfy service level requirements. It facilitates the mobile service provider to decide on the number of channel for a MSC to satisfy the policy of service level requirements.

Glossary of Abbreviations

ACST Average call setup time

AuC Authentication Center

BSC Base station controller

BSS Base station subsystem

BTS Base transceiver station

EIR Equipment Identity Register

FBSD Fixed block of seven days

GMSC Gateway Mobile Switching Centre

GSM Global Systems for Mobile Communication

GTT Global transition table

HLR Home location register

HR Hit rate

ICCID Integrated circuit card ID

IMEI International Mobile Equipment Identity

IMSI International Mobile Subscriber Identity

ISDN Integrated service digital network

LA Location area

LAI Location area identity

© Springer Nature Singapore Pte Ltd. 2017
N. Mallikharjuna Rao and M. Muniratnam Naidu, *Sliding Window Algorithm for Mobile Communication Networks*, https://doi.org/10.1007/978-981-10-8473-7

MAP Mobile application part

MCC Mobile country code

MM Mobility management

MNC Mobile network code

MS Mobile subscriber

MSC Mobile Switching Centre

MSIN Mobile station identification

MSISDN Mobile Station International Subscriber Directory Number

MSRN Mobile Subscriber Roaming Number

MTSO Mobile telephone switching office

NSS Network switching subsystem

OMC Operations and maintenance center

OSS Operation support subsystem

PIN Personal identification number

PSTN Public switched telephone network

RA Routing area

RAI Routing area identity

RM Resource management

SID Subscriber ID's

SIM Subscriber information module

SS7 Signaling system no 7

SSP Switching service point

STP Signal transfer point

SWSSD Sliding window of size seven days

TDMA Time division multiple access

TLDN Temporary location directory number

TMSI Temporary mobile subscriber identity

TP Throughput

VLR Visitor location register

WS Window size

References

1. Elnahas, A., Adly, N: Location management techniques for mobile systems. Inf. Sci. **130**, 1–22 (2000) (Elsevier)
2. Panda, R.: Mobile and Personal Communication Systems and Services. Prentice Hall India (2000)
3. Agrawal, D., Zhang, Q.-A.: Introduction to Wireless and Mobile Systems. Cole and Brooks Publishers, Pacific Grove, CA (2003)
4. Lee, J.-W.: Mobility management using frequently visited location database. In: International Conference on Multimedia and Ubiquitous Engineering, IEEE (2007)
5. Mao, Z., Douligeris, C.: A distributed database architecture for global roaming in next-generation mobile networks. IEEE/ACM Transaction on Networking, **12**(1) (2004)
6. Rahnema, M.: Overview of the GSM system and protocols architecture. IEEE Communications Magazine, April (1993)
7. Lee, W.C.Y.: Mobile Cellular Communications: Analog and Digital Systems, 2nd edn. McGraw-Hill Inc. ISBN: 0-0703-8089-9 (1995)
8. Scourias, J.: Overview of GSM: The Global System for Mobile Communications (1996)
9. Jain, R.: Reduced traffic impacts of PCS using hierarchical user location databases. In: Proceeding of IEEE International Conference on Communication (1996)
10. Jain, R., Lin, Y.B., Lo, C., Mohan, S.: A caching strategy to reduce network impacts of PCS. IEEE Selected Area Comm. **12**(8), 1334–1444 (1994)
11. Ho, J.S.M., Akyildiz, I.F.: Local anchor scheme for reducing signaling cost in PCS networks. IEEE/ACM Trans. Networking **4**(5), 709–725 (1996)
12. Mao, Z., Douligeris, C.: Performance evaluation of location information distribution strategies for mobility tracking. In: 7th International Conference on Parallel and Distributed Systems, IEEE (2000)
13. Mao, Z., Douligeris, C.: A distributed database architecture for global roaming in next generation mobile networks. IEEE/ACM Trans. Networking **12**(1), 146–160 (2004)
14. Huh, Y., Chot, M., Kim, C.: A delayed location registration procedure for wireless mobile communication networks. IEICE Trans. Commun. **E84-B**(10), 2805–2815 (2001)
15. Oh, S.-J.: The location management scheme using mobility information of mobile users in wireless mobile networks. In: Proceedings of 2003 International Conference on Computer Networks and Mobile Computing, IEEE (2003)
16. Luo, Y., Pan, Y., Li, J., Xiao, Y., Lin, X.: A new overflows Replacement policy for efficient location management in mobile networks. In: 23rd International Conference on Distributed Computing Systems Workshop, p. 780, ISBN: 0-7695-1921-0 (2003)
17. Giner, V.C., Garcia-Escalle, P.: On the fractional movement-distance based scheme for PCS location management with selective paging. Wirel. Syst. Mobility Next Gener. Internet **3427**, 202–218 (2005)

© Springer Nature Singapore Pte Ltd. 2017
N. Mallikharjuna Rao and M. Muniratnam Naidu, *Sliding Window Algorithm for Mobile Communication Networks*, https://doi.org/10.1007/978-981-10-8473-7